Oxford KS3 Science

Activate 2

Question • Progress • Succeed

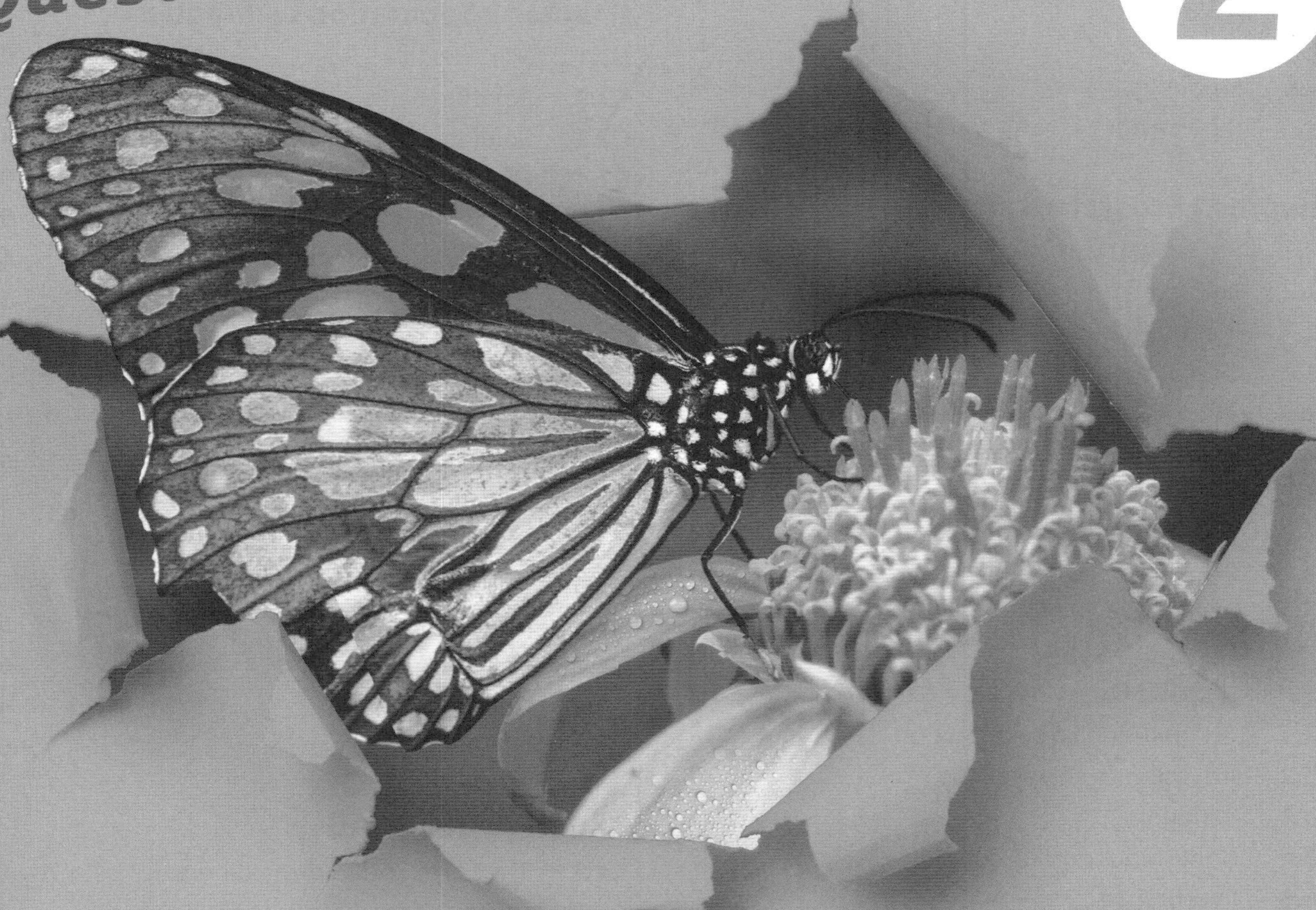

Intervention Workbook
Foundation

Jon Clarke
Philippa Gardom Hulme
Jo Locke

Assessment Editor
Dr Andrew Chandler-Grevatt

OXFORD
UNIVERSITY PRESS

Contents

Physics P2

Chapter 1 Electricity and magnetism

Chapter 2 Energy

Chapter 3 Motion and pressure

Introduction

Welcome to your *Activate* 2 Workbook. This Workbook contains lots of practice questions and activities to help you to progress through the course.

Each chapter from the *Activate* 2 Student Book is covered and includes a summary of all the content you need to know. Answers to all of the questions are in the back of the Workbook so you will be able to see how well you have answered them.

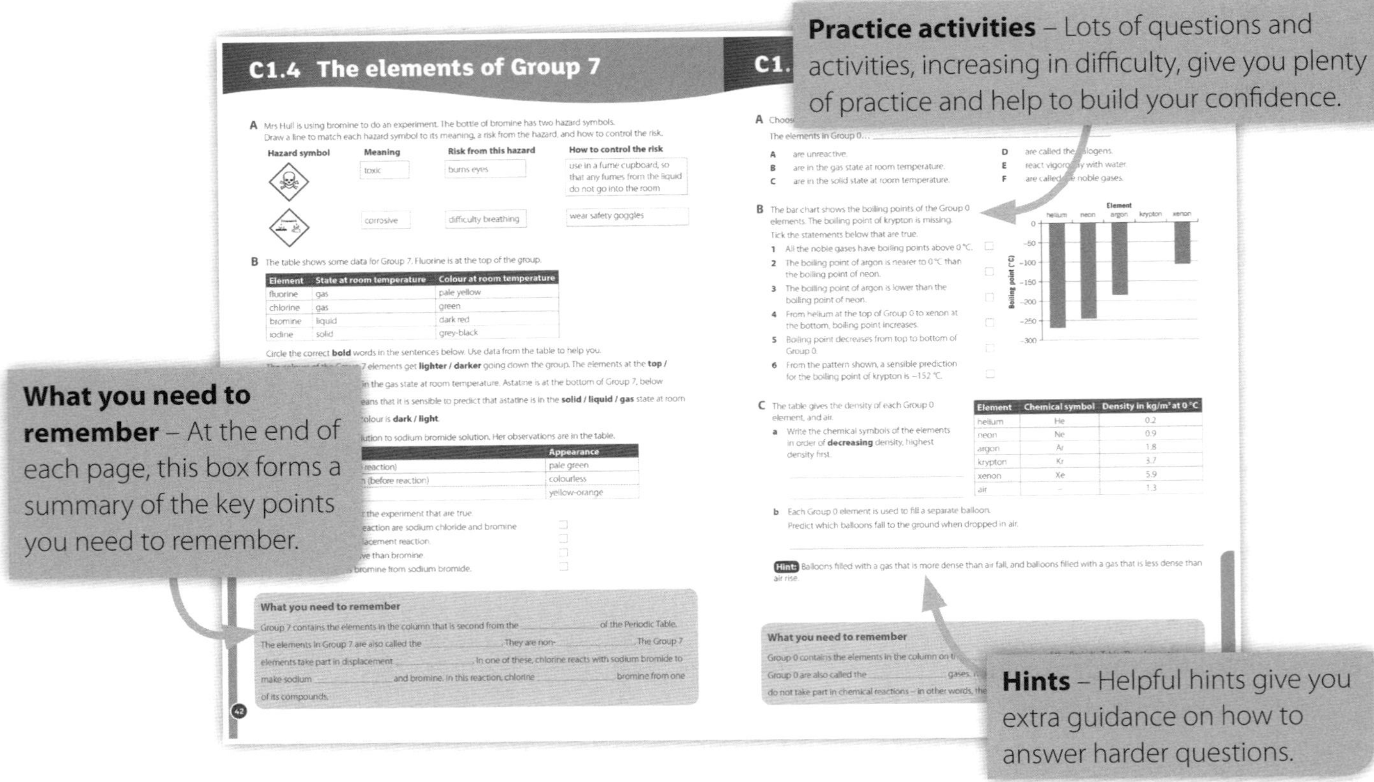

Practice activities – Lots of questions and activities, increasing in difficulty, give you plenty of practice and help to build your confidence.

What you need to remember – At the end of each page, this box forms a summary of the key points you need to remember.

Hints – Helpful hints give you extra guidance on how to answer harder questions.

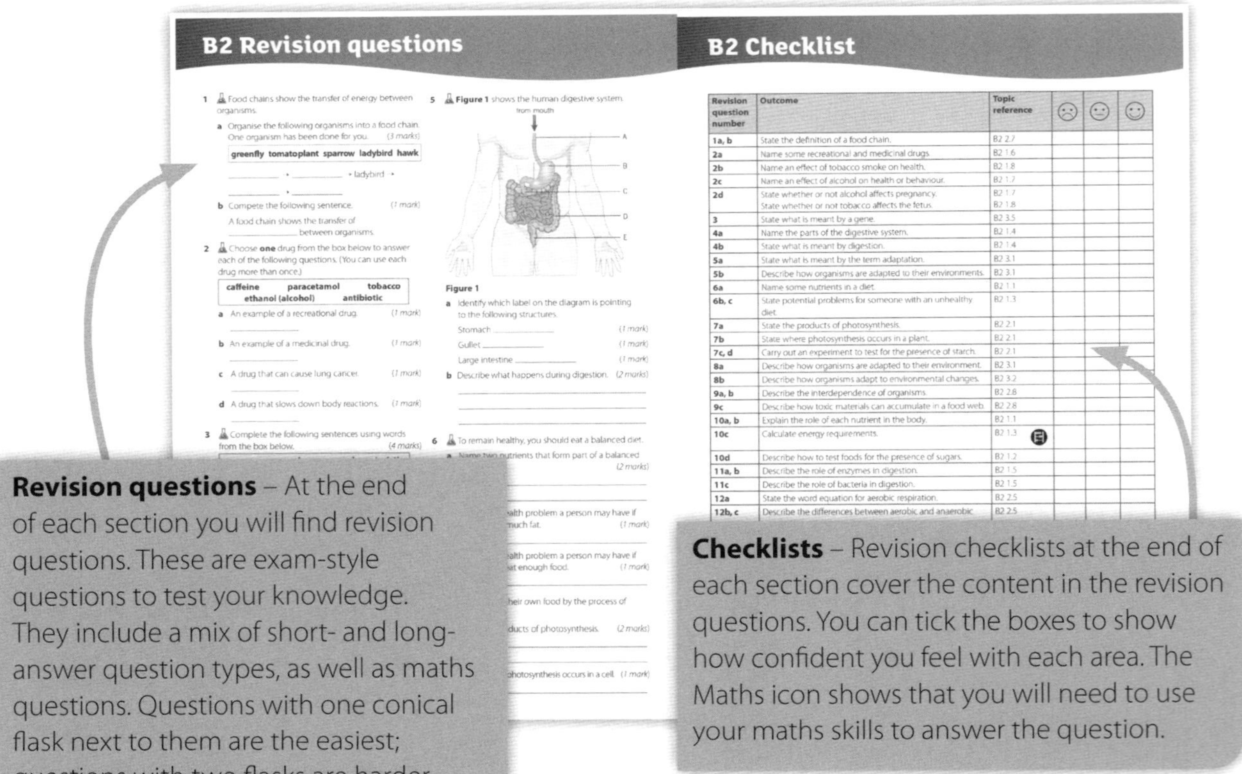

Revision questions – At the end of each section you will find revision questions. These are exam-style questions to test your knowledge. They include a mix of short- and long-answer question types, as well as maths questions. Questions with one conical flask next to them are the easiest; questions with two flasks are harder.

Checklists – Revision checklists at the end of each section cover the content in the revision questions. You can tick the boxes to show how confident you feel with each area. The Maths icon shows that you will need to use your maths skills to answer the question.

Pinchpoints

A Pinchpoint is an idea or concept in science that can be challenging to learn. It is often difficult to say *why* these ideas are challenging to learn. The Pinchpoint intervention question at the end of each chapter focuses on a challenging idea from within the chapter. By answering the Pinchpoint question you will see whether you understand the concept or whether you have gone wrong. By doing the follow-up activity you will find out why you made the mistake and how to correct it.

Pinchpoint question – The Pinchpoint question is about a difficult concept from the chapter that students often get wrong. You should answer the Pinchpoint question and one follow-up activity. The Pinchpoint is multiple choice; answer the question by choosing a letter and then do the follow-up activity with the same letter.

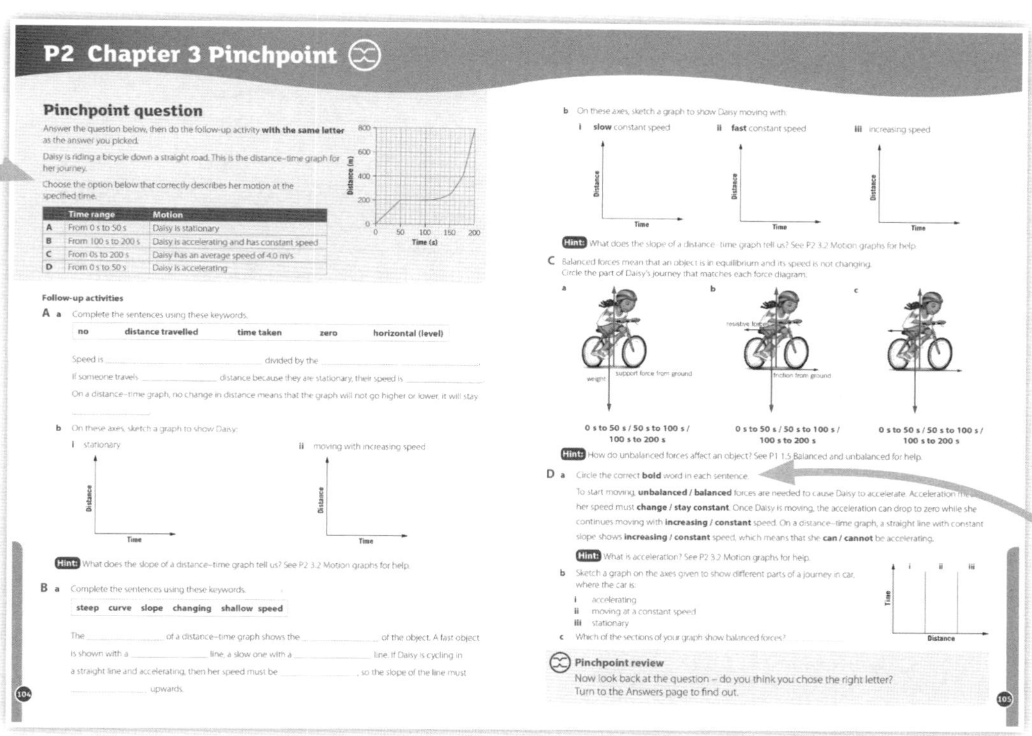

Pinchpoint follow-up – The follow-up activities will help you to better understand the difficult concept. If you got the Pinchpoint question right, the follow-up will develop your understanding further. If you got the Pinchpoint wrong, it will help you to see why you went wrong, and how to get it right next time.

B1.1 Nutrients

A From the list below, circle the **nutrients** your body needs.

| lipid | oxygen | blood | protein | vitamin | bread | carbohydrate |

B Choose **one** food type from the box below to complete each row of the table.

| fruit | pasta | fish | butter |

Good source of ……..	Food
Protein	
Vitamins and minerals	
Carbohydrate	
Lipid	

C Draw a line to match each nutrient with its role in the body.

carbohydrate		growth and repair of body tissues
lipid		main source of energy
protein		needed in all cells and body fluids
water		store of energy, keep you warm, protect organs
vitamin		needed in tiny amounts to keep you healthy

D Circle **true** or **false** for the following statements about fibre.

1 It is a nutrient. **true / false**

2 It adds bulk to help keep food moving through gut. **true / false**

3 It helps prevent constipation. **true / false**

4 It is found in animal products like milk. **true / false**

What you need to remember

To remain healthy you must eat a _______________ diet. This means eating food containing the right _______________ in the right amounts. These include _______________ and _______________ which give you energy, _______________ for growth and repair, _______________ and _______________ to keep you healthy, and water and _______________ to keep the food moving through your gut.

"

B1.2 Food tests

A Many food tests involve adding a solution to the food sample.
Tick the change below that is used to show which nutrients the food contains.

The solution evaporates ☐

A solid is formed ☐

The solution changes colour ☐

A gas is given off ☐

B The following pieces of equipment are used in food tests.

> **water bath   pestle and mortar   filter paper   pipette**

Complete the table to match the piece of equipment with its use.

Equipment	Use
	Grind up food into small pieces
	Transfer small amounts of liquid
	Remove solid material from solution
	Heat solution to a specific temperature

C Circle the correct **bold** words in the sentences below to describe how you can test for lipids in a solid piece of food.

Rub some food onto a piece of **filter paper / cardboard**.

Hold the paper **under running water / to the light**.

If the paper **goes translucent / turns blue** it contains lipids.

D You can test foods to find out what food molecules they contain.

Draw lines to connect the food being tested to the chemical used and the result if the molecule is present

Food molecule	Chemical to use	Result if food being tested contains the molecule
starch	ethanol	blue-black colour
sugar	blue Benedict's solution	purple colour
protein	orange iodine solution	cloudy, white layer
lipid	blue copper sulfate and sodium hydroxide solution	brick-red colour

What you need to remember

Scientists use _________________________ to find out which nutrients are present in a food product.

_________________ turns blue-black when _________________ is present. Benedict's solution turns orange-

_________________ if sugar is present. A solution of copper sulfate and sodium hydroxide solution will turn

_________________ if _________________ is present. Ethanol will turn _________________ if lipids are present.

B1.3 Unhealthy diet

A Circle the correct **bold** words in the sentences below.

Your body needs **energy / light** to function properly.

The energy contained in food is measured in **grams / joules**.

The amount of energy everybody needs is the **same / different**.

The more exercise you do, the **more / less** energy your body requires.

B The following people need different amounts of energy each day. Put them in order, from those who need the least to the most energy.

Correct order ☐ ☐ ☐ ☐

1 adult male builder

2 teenager

3 5-year-old child

4 adult male IT technician

C If you are malnourished, you are eating the wrong amount or wrong type of food.
Draw a line to match each health issue with its dietary cause.

Obesity	Eating foods lacking in vitamins or minerals
Deficiency	Eating too much food or too many fatty foods
Starvation	Eating too little food

D The list shows some symptoms of an unbalanced diet. Sort the symptoms into the correct column to show whether they are more likely to occur in a person who is obese or underweight.

lack of energy	diabetes	heart disease	poor immune system

Obese	Underweight

What you need to remember

Eating the wrong amount or wrong types of food is called ____________.

If the energy in the food you eat is less than the energy you use, you will lose body mass and become

____________. You are also likely to not take in the correct amount of a vitamin or mineral. This is

called a ____________ and can make you ill. Not eating enough food for prolonged periods is

called ____________.

If you take in more energy than you use by eating too much, you will gain body mass as ____________,

which is stored under the skin. Extremely overweight people are said to be ____________.

B1.4 Digestive system

A Circle the correct **bold** words in the sentences below to describe what happens during digestion.

During digestion, **large / small** molecules such as proteins are **broken down / joined together** into **large / small** molecules.

B Label the diagram of the digestive system using the key terms from the box below.

stomach	rectum	small intestine	gullet	large intestine

C Complete the table below to show the where the different parts of digestion occur.

stomach	liver	small intestine	gullet	large intestine

Digestive system organ	What happens there
	nutrient molecules are absorbed into blood
	undigested food (feces) leaves the body
	food is mixed with digestive juices and acids
	muscular tube that squeezes food from the mouth into the stomach
	water passes into body and feces form

What you need to remember

The group of organs which work together to break down food is called the _______________________. Food enters the mouth and travels down your _______________ into your _________________. Here it is mixed with _______________ and digestive juices. As a result of _________________, small molecules of nutrients are produced which pass through the _______________ intestine into the blood. Water passes back into the body in the _______________ intestine leaving undigested food called feces. This is stored in the _______________ until it leaves the body through the _______________.

B1.5 Bacteria and enzymes in digestion

A Enzymes are involved in digestion. Circle the enzymes in the list below.

| bile | bacteria | protease | carbohydrase | glycerol | probiotic | lipase |

B Where in your digestive system do helpful bacteria make vitamins?

Mouth ☐

Stomach ☐

Small intestine ☐

Large intestine ☐

C Circle **true** or **false** for the following statements about enzymes.

1 They are made of lipids.   **true / false**
2 They speed up digestion.   **true / false**
3 They are known as biological catalysts.   **true / false**
4 They are used up during a reaction.   **true / false**

D Each type of enzyme is involved in a different reaction.
Draw a line to match each type of enzyme to the molecule it breaks down and the molecules that are produced.

Enzyme	**Molecule it breaks down**	**Molecules produced**
Carbohydrase	Lipid	Sugar
Protease	Carbohydrate	Fatty acids and glycerol
Lipase	Protein	Amino acid

What you need to remember

Some _______________ living in your large intestine help you to remain healthy by making _____________.

Special proteins called _________________ help speed up digestion without being used up. They are a type

of _______________. There are three main types – _________________ which breaks down carbohydrate

molecules into _________________ molecules, _________________ which breaks down _________________

into amino acids and _________________ which breaks down lipid molecules into fatty acids and _____________.

To help further with lipid digestion, _________________ breaks the lipids into smaller droplets that are easier for

the enzymes to work on.

B1.6 Drugs

A Drugs are chemicals which affect the way your body works.
Sort the following list of drugs into those which are medicinal and those which are recreational.

ecstasy	caffeine	antibiotic	alcohol	paracetamol	tobacco	ibuprofen

Medicinal	Recreational

B Circle **medicinal** or **recreational** for the following statements about drugs.

1 They have health benefits.   **medicinal / recreational**

2 They are taken for enjoyment.   **medicinal / recreational**

3 Many are illegal.   **medicinal / recreational**

4 They are prescribed by a doctor.   **medicinal / recreational**

C Draw a line to match each recreational drug with its effect on health.

alcohol		speeds up the nervous system
tobacco		significantly increases risk of lung cancer and heart disease
caffeine		slows down the nervous system and damages the liver

D Identify some common withdrawal symptoms an addict may experience if they try to stop taking a drug.
Tick as many boxes as you need.

Headaches ☐

Anxiety ☐

Happiness ☐

Excess sweating ☐

What you need to remember

Chemicals that affect the way your body works are called ________________. ________________ drugs are

taken for enjoyment whereas ________________ drugs benefit your health. If you regularly take a drug, you

may develop an ________________. If you then try to stop taking the drug, you may suffer from unpleasant

________________ ________________, which make it harder to give up.

B1.7 Alcohol

A Drinking alcohol has several effects on the body.
Write these effects in the correct box on the diagram opposite:

unconsciousness

feeling relaxed and happy

difficulty walking and talking (drunk)

death

no alcohol

INCREASING INTAKE

excessive alcohol

How a person would be affected

B Tick the **two** organs in your body alcohol is most likely to damage.

Lungs ☐ Liver ☐

Brain ☐ Heart ☐

C Circle the correct **bold** words in the sentences below about alcohol.

Alcohol contains the drug **caffeine / ethanol**. This is absorbed into your **lungs / bloodstream**, and then travels to the brain affecting your **nervous / digestive** system. It is called a **stimulant / depressant** because it **slows down / speeds up** your reactions.

D Circle **true** or **false** for the following statements about alcohol and pregnancy. Drinking alcohol…

1	increases the risk of miscarriage	**true / false**
2	reduces the birth weight of babies	**true / false**
3	increases the amount of sperm produced	**true / false**
4	can damage the fetus's brain, causing learning difficulties	**true / false**

What you need to remember

Alcoholic drinks contain the drug _______________. This acts on the _______________ system and slows down body reactions; it is called a _______________. Drinking too much alcohol can result in _______________ and brain damage. Different alcoholic drinks contain different amounts of alcohol. 10 ml of alcohol is known as one _______________ of alcohol. The government recommends that adults drink less than 2–3 units a day to remain healthy. People who are addicted to alcohol are known as _______________.

B1.8 Smoking

A Smoking increases your chances of developing many diseases.

Circle **three** diseases people are more likely to develop if they smoke.

diabetes	lung cancer	heart attack	AIDS	stroke	measles

B Choose which of the following components of tobacco smoke are being described in the sentences below. You can use each word more than once.

tar	carbon monoxide	nicotine

Stimulant which speeds up the nervous system: _________________

Sticky black material that collects in lungs: _________________

Addictive chemical: _________________

Contains chemicals which can cause cancer: _________________

Gas which stops blood from carrying as much oxygen as it should: _________________

C Tick **two** boxes to show the risks of smoking in pregnancy:

Baby being obese ☐

Low birth weight baby ☐

Miscarriage during pregnancy ☐

Baby suffering from Down's syndrome ☐

Pinchpoint question

Answer the question below, then do the follow-up activity **with the same letter** as the answer you picked.

Digestion begins in your mouth.

Read the statements below and choose which one correctly describes the action of the enzymes present in saliva.

A The enzymes live in cells and speed up a reaction before being used up.

B The enzymes break down carbohydrates into amino acids.

C Saliva contains protease enzymes which break down protein molecules.

D The enzymes break down carbohydrate molecules into sugar molecules.

Follow-up activities

A Write down **four** correct sentences about enzymes using the sentence starters and endings below. You should use one starter twice to make the fourth sentence.

Enzymes are catalysts. This means they…	…are not used up in a reaction.
Enzymes are made of…	…respire.
Enzymes are not living as they cannot…	…speed up a reaction.
	…protein molecules.

Hint: Remember all living organisms have to be able to perform MRS GREN. See B2 1.5 Bacteria and enzymes in digestion for help.

B Complete the diagrams below to show how the enzymes break large molecules down into smaller molecules during digestion. Fill in all the missing names for the reactants, enzymes, and products.

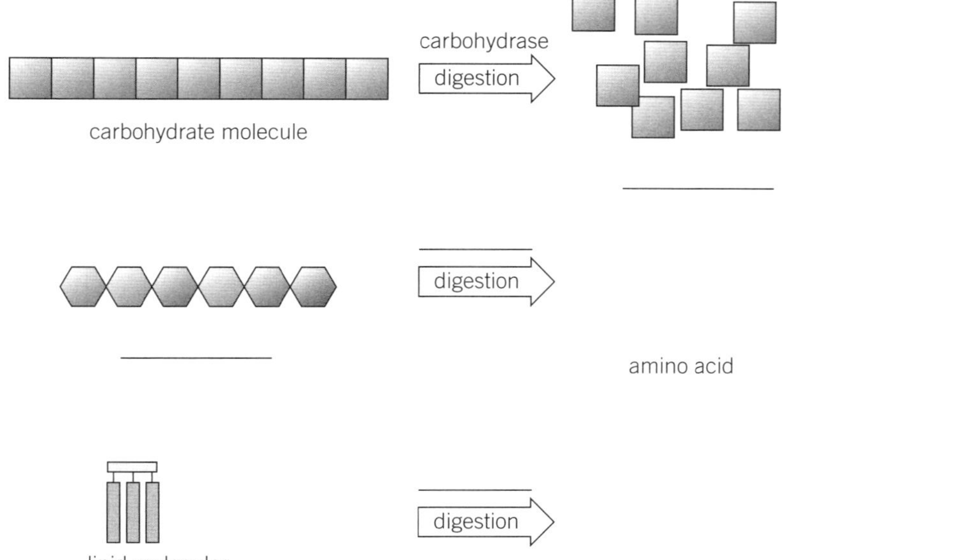

Hint: Look at the start of each enzyme's name. It gives you a clue to what it breaks down. See B2 1.5 Bacteria and enzymes in digestion for help.

C Complete the following table to show where enzymes are found in the body. Add a tick for each place the enzyme is found.

Enzyme	Mouth	Stomach	Small intestine
Carbohydrase			
Protease			
Lipase			

Hint: There is only one type of enzyme found in the mouth. It breaks down large carbohydrate molecules into sugars. See B2 1.5 Bacteria and enzymes in digestion for help.

D There are two main types of washing powder: biological and non-biological powders. Biological washing powders contain enzymes as well as detergents; non-biological powders do not contain enzymes.

Explain why enzymes are added to washing powder to help get clothes clean. Include a specific example to help in your explanation.

____

____

____

____

____

____

Hint: Think about the cause of many of the 'dirty' marks on clothes. See B2 1.5 Bacteria and enzymes in digestion for help.

Pinchpoint review

Now look back at the question – do you think you chose the right letter?
Turn to the Answers page to find out.

B2.1 Photosynthesis

A Plants make their own food through the process of photosynthesis. Tick where photosynthesis takes place.

Nucleus ☐

Cytoplasm ☐

Chloroplast ☐

Mitochondria ☐

B Tick the **two** products of photosynthesis.

Glucose ☐

Fat ☐

Oxygen ☐

Carbon dioxide ☐

C Use the words below to complete the word equation for photosynthesis.

oxygen	water	light

____

Carbon dioxide + _________________ ⟶ glucose + _________________

D Draw a line to match each substance needed in photosynthesis with how it enters the plant.

Carbon dioxide	Absorbed by chlorophyll in chloroplasts
Water	Enters through tiny holes on the underside of the leaf
Light	Diffuses into root hair cells

What you need to remember

Plants and _________________ are called _________________ because they make their own food by the process

of _________________. Animals are called _________________ as they have to eat other organisms to survive.

During photosynthesis, carbon dioxide and _________________ are converted into oxygen and _________________

using energy from the Sun. This light energy is absorbed by _________________ in chloroplasts.

B2.2 Leaves

A Label the diagram of the cross section through a leaf using the key words in the box below.

> **waxy layer**   **palisade layer**   **spongy layer**   **chloroplast**
>
> **air space**   **stoma**   **guard cell**

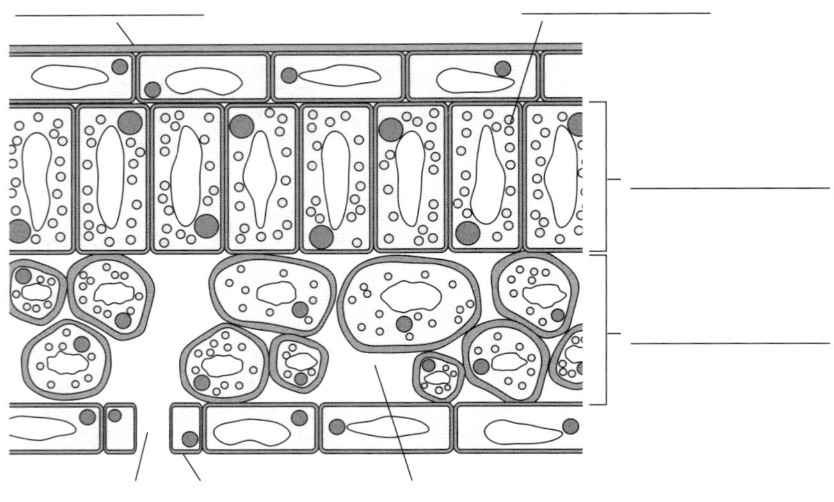

B Draw a line to match each main component in a leaf with its function.

Chloroplast	Open and close stomata
Stomata	Contains chlorophyll to trap sunlight
Guard cells	Transport water to cells in leaf
Veins	Allow gases to diffuse into and out of leaf
Waxy layer	Reduces water loss through evaporation

C Circle **true** or **false** for the following statements about the palisade layer.

1 Found at the bottom of the leaf   **true / false**
2 Main site of photosynthesis   **true / false**
3 Contains many air spaces   **true / false**
4 Contains cells packed with chloroplasts   **true / false**

What you need to remember

Photosynthesis in a plant mainly takes place in the _______________, though a small amount occurs in the stems.

The underneath of a leaf contains tiny holes called _______________ which allow _______________ _______________

to diffuse into the leaf and _______________ to diffuse out. Water is carried to the leaf in the _______________.

Most photosynthesis occurs in the cells of the _______________ layer as most light reaches this layer. Therefore,

these cells are full of _______________.

B2.3 Plant minerals

A To remain healthy, plants need to take in minerals. Circle the minerals in the list below.

water	nitrates	glucose	oxygen	potassium

B To ensure plants have enough minerals and do not suffer from a deficiency, some farmers add man-made chemicals to the soil plants are growing in. Tick the chemicals added.

Anti-fungals ☐

NPK fertilisers ☐

Manure ☐

Compost ☐

C Draw a line to match each mineral to its use in the plant, and describe the plant's appearance if this mineral is deficient.

Mineral	Function	Deficiency symptoms
Nitrate	Making chlorophyll	
Phosphate	Healthy growth	
Potassium	Healthy roots	
Magnesium	Healthy leaves and flowers	

D Circle the correct **bold** words in the sentences below to explain how nitrates are taken in by plants and involved in growth.

Plants get the minerals they need from the **air / soil**. The minerals are dissolved in the **air / soil**. Minerals are absorbed by the plant's **stomata / root hair cells** and are then transported around the plant by the **phloem / xylem**. Nitrates are involved in making **amino acids / sugars**. These join together to form **carbohydrates / proteins**.

What you need to remember

To stay healthy, plants need to absorb _________________ from the soil. For healthy growth, plants need four minerals – _________________ to make chlorophyll, _________________ for healthy leaves and flowers, _________________ for healthy growth and _________________ for healthy roots. If a plant does not get enough of a mineral it is said to have a _________________ and will not grow properly. To prevent this occurring, farmers add chemicals called _________________ to the soil.

B2.4 Chemosynthesis

A Some organisms use chemosynthesis to make food. Tick which type of organism performs this chemical reaction.

Plants ☐

Animals ☐

Bacteria ☐

B Some chemosynthetic bacteria live in tube worms and both organisms benefit.
What is this relationship called? Tick the correct box.

Antagonistic ☐

Mutualistic ☐

Photosynthetic ☐

C Complete the sentences on chemosynthesis using the words in the box below.

| light | glucose | oxygen | starch | a chemical | carbon dioxide | dark |

The source of energy for chemosynthesis is ____________________.

The product of chemosynthesis is ____________________.

A reactant often used in chemosynthesis is ____________________.

Chemosynthesis takes place in organisms that live in the ____________________.

D Bacteria that perform chemosynthesis are called chemosynthetic bacteria.

Draw a line to match each type of bacteria with where it is found and the chemical it uses.

| Sulfur bacteria | | Near volcanic vents at the bottom of the sea | | Nitrogen compounds |

| Nitrogen bacteria | | Plant roots | | Hydrogen sulfide |

What you need to remember

Some species of ____________________ make their own food by the process of ____________________. They use the

energy released by ____________________ reactions to make ____________________. One example of chemosynthetic

bacteria is ____________________ bacteria living at the bottom of the ____________________ near volcanic vents.

B2.5 Aerobic respiration

A Tick the box which names the process where energy is released in cells.

Photosynthesis ☐   Respiration ☐

Chemosynthesis ☐   Digestion ☐

B Tick the **two** reactants needed for aerobic respiration.

Glucose ☐   Water ☐

Oxygen ☐   Carbon dioxide ☐

C Complete the word equation for aerobic respiration using words from the box below.

oxygen	**water**	**light**	**carbon dioxide**	**starch**

glucose + _________________ ⟶ water + _________________ (+ energy)

D Aerobic respiration takes place in one part of the cell.
On the diagram of an animal cell below, label this part with its name.

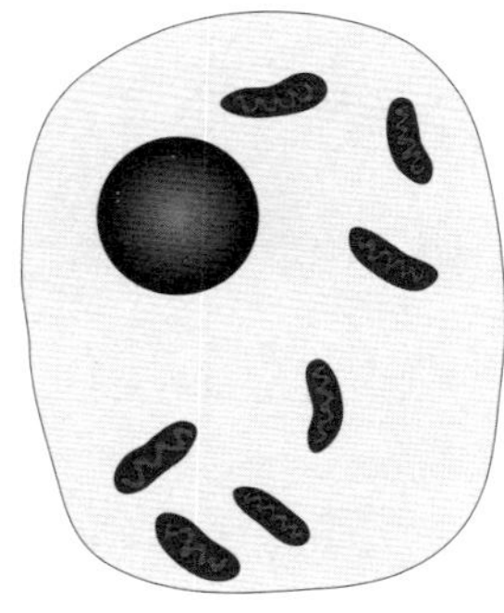

B2.6 Anaerobic respiration

A When there is no oxygen present, animals respire anaerobically.
Tick the product of anaerobic respiration in animals.

Glucose ☐　　　　Lactic acid ☐

Ethanol ☐　　　　Carbon dioxide ☐

B Which of the following word equations correctly shows anaerobic respiration in animal cells?

glucose → lactic acid + carbon dioxide ☐

glucose → lactic acid ☐

glucose → ethanol + carbon dioxide ☐

glucose → water + carbon dioxide ☐

C The table shows some statements about respiration.
Tick **one** column in each row to show which statements are true for each type of respiration.

	✓ if true for aerobic respiration	✓ if true for anaerobic respiration
Glucose is a reactant		
Oxygen is a reactant		
Carbon dioxide is produced		
Lactic acid is produced		
Water is produced		
Transfers more energy per glucose molecule		

D Circle **true** or **false** for the following statements about anaerobic respiration.

1 Lactic acid produced can cause muscle cramp.　　　　**true / false**

2 Carbon dioxide is used to break down lactic acid.　　　　**true / false**

3 When yeast respires anaerobically, it produces lactic acid.　　　　**true / false**

4 The word equation for fermentation is:
glucose → ethanol + carbon dioxide (+ energy)　　　　**true / false**

What you need to remember

When your body respires without oxygen it is called ________________________________. This produces
________________________________ which can build up in your muscles and cause cramp. To break down
the acid, you have to breathe in extra oxygen. This is called an ________________________________.

Microorganisms such as yeast carry out a type of anaerobic respiration called ________________. In this reaction,
carbon dioxide and ________________ are produced.

B2.7 Food chains and webs

A Draw a line to match each term with its definition.

Food chain	An organism that makes its own food
Food web	A diagram that shows the transfer or energy between organisms
Producer	An organism that eats other organisms to gain energy
Consumer	A diagram showing linked food chains

B This food web shows the relationships between organisms that live in Africa.

Give the name of an organism from the food web that is a:

a Producer ____

b Herbivore ____

c Carnivore ____

d Predator ____

e Prey ____

C Using information from the food web, draw a food chain with four links.

____________ ⟶ ____________ ⟶ ____________ ⟶ ____________

What you need to remember

A _______________________________ is a diagram that shows the transfer of ______________ between organisms.

The first organism is always a ______________. It transfers energy from the Sun into glucose by ______________.
The other organisms in the chain are ______________. They eat other organisms to gain energy. An animal that

is eaten by another organism is called a ______________ organism. The animal that eats it is called a

______________. Most animals have more than one food source. This can be shown on a ______________

______________. These diagrams show a set of ______________ food chains.

A Complete the sentences below using the words from the box.

depend	population	interdependence	increase	shelter	species

The organisms within a food chain _____________________ on each other for survival. This is called

_____________________. For example, animals need plants for food and _____________________.

The number of plants or animals of the same _____________________ living in an area is called a

_____________________. The size of one population of organisms can affect another. For example, if the number

of lettuces increases, the number of rabbits will _____________________ as more food is available for survival.

B Tick the term which describes the build-up of toxic chemicals in organisms through a food chain.

Bioaccumulation ☐

Symbiosis ☐

Chemosynthesis ☐

C The diagram opposite shows part of a woodland food web.

There is an outbreak of a disease that kills only ladybirds. Use the food web to explain how this affects other organisms in the woodland.

Circle the correct **bold** word and complete each sentence.

a The number of ladybirds will **increase / decrease** because

____

____

b The number of aphids will **increase / decrease** because

____

____

What you need to remember

Living organisms depend on other organisms to survive. This is called _______________. The number of organisms

of a particular species in an area is called a _______________. If the size of one population increases, it can change

the size of another population. For example, if a producer population increases, the consumer population

may _______________.

Toxic chemicals can build up in the organisms in a food chain. This is called _______________________________.

B2.9 Ecosystems

A Draw a line to match each key term to its definition.

ecosystem	particular place or role that an organism has in an ecosystem
niche	different types of organism living in the same place at the same time
co-exist	the living organisms in a particular area and the habitat in which they live
habitat	the organisms in an ecosystem
community	the area where an organism lives

B A scientist wanted to investigate the different types of plant found in a meadow. Which piece of equipment should he use to sample the ecosystem?

Pooter ☐

Sweep net ☐

Quadrat ☐

Pitfall trap ☐

C A group of students looked at the organisms present in an oak tree ecosystem. They found birds, ants, squirrels, woodlice, and slugs living on the oak trees.

a Complete the sentences below to describe differences in the organisms' niches.

Birds and squirrels both live in the tree canopy. They can co-exist as they have different _______________

_______________ .

Squirrels and woodlice can co-exist as they are found in different _______________.

b In the sentences above, circle the habitat and underline the community.

Pinchpoint question

Answer the question below, then do the follow-up activity **with the same letter** as the answer you picked.

Which of the following statements best describes the process of aerobic respiration?

A It is represented by the word equation: carbon dioxide + water → glucose + oxygen (+ energy)

B A process that occurs in only animal cells to transfer energy for movement.

C A chemical reaction where glucose reacts with oxygen to release energy and other waste products.

D Taking in oxygen through your lungs and releasing carbon dioxide.

Follow-up activities

A The following sentences describe what happens in aerobic respiration.
Read the sentences then complete the activities below.

Your body needs energy for everything it does.

You get your energy from organic molecules in the food you eat.

To release the energy stored in food, glucose reacts with oxygen in a reaction called aerobic respiration.

This transfers energy to your cells.

As a result of the reaction, carbon dioxide and water are produced.

a Underline the **two** reactants of aerobic respiration.

b Circle the **two** waste products of aerobic respiration.

c Complete the word equation for respiration.

___________________ + ___________________ ⟶ ___________________ + ___________________

Hint: Reactants are the starting substances in a chemical reaction. See C1 3.2 Word equations and B2 2.5 Aerobic respiration for help.

B **a** Complete the following table using the terms **yes**, **no**, or **yes – some species** to show what types of chemical reaction occur in each of the following organisms.

Organism	Aerobic respiration	Photosynthesis
Animal		
Plant		
Microorganism		

b Complete the following sentences.

All living organisms r ________________. to transfer energy from glucose. This can be used for

m ________________. and g ________________.

Green organisms, such as p ________________. and some b ________________, are able to

p ________________ to transfer light energy. This is used to produce g ________________

molecules for the organism.

Hint: Remember all living organisms have to be able to perform MRS GREN. Can you remember what each letter in this mnemonic stands for? See B2 2.1 Photosynthesis and B2 2.5 Aerobic respiration for help.

C The energy released during respiration supplies all the energy needed for living processes in a cell. Explain why the following cells need energy:

a Muscle cell

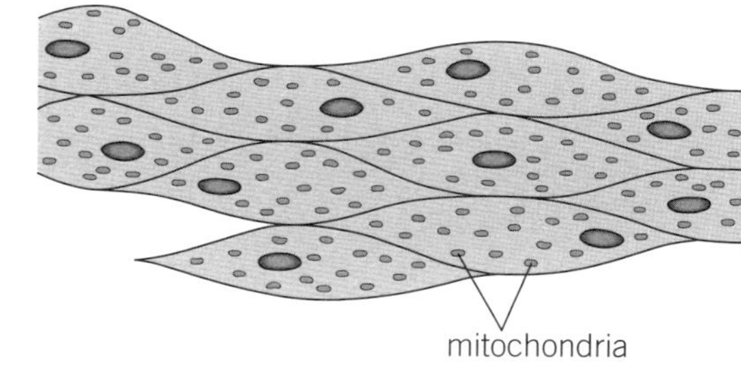

____

____

____

____

____

b Sperm cell

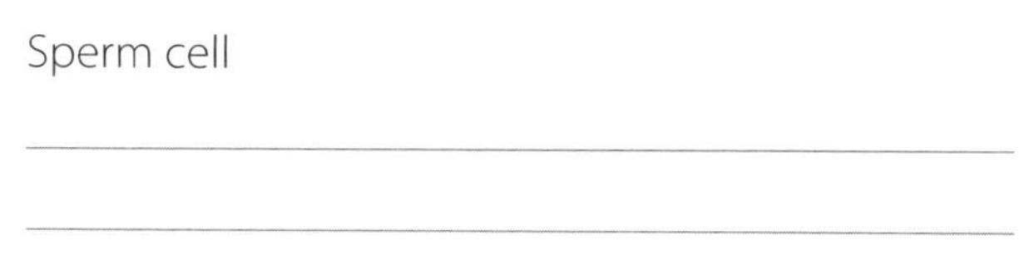

____

____

____

____

____

c The unicellular organism euglena

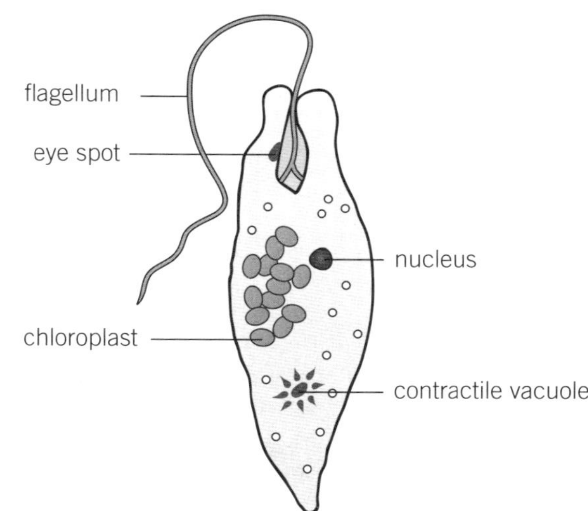

____

____

____

____

____

Hint: Think about the main function of each cell. See B2 2.5 Aerobic respiration and B1 1.5 unicellular organisms for help.

D Circle the correct **bold** terms in the sentences below to describe how the substances needed for respiration enter the body, and the waste products produced leave the body.

Glycerol / glucose is a carbohydrate found in food. These small molecules are released during **digestion / photosynthesis**.

Glucose molecules are absorbed by the wall of the **large / small** intestine and pass into your blood.

Glucose **diffuses / dissolves** into the cells that need it for respiration.

Oxygen is taken into your lungs when you **inhale / exhale**. This is also known as **breathing / respiration**.

It diffuses into your blood through the **amoeba / alveoli** and is carried to the cells which need it for respiration.

Carbon dioxide / nitrogen is produced as a result of respiration. If this builds up in the body it can cause harm.

Carbon dioxide diffuses into the blood and is carried to the lungs where it is **inhaled / exhaled**.

Hint: You breathe to allow gas exchange to take place in the lungs. Where does respiration occur in the body? See B2 2.5 aerobic respiration for help.

Pinchpoint review

Now look back at the question – do you think you chose the right letter?
Turn to the Answers page to find out.

B3.1 Competition and adaptation

A Plants and animals have special characteristics to help them survive.
What is the term used to describe these special characteristics? Tick the correct box.

competition ☐

adaptations ☐

aids ☐

B Organisms compete with each other for resources. Circle the resources that animals compete for.

food	**space to hunt**	**water**	**light**	**space to grow**	**mates**

C Plants also compete with each other for resources.
Draw a line to match each resource with why it is needed.

Light	To make chemicals needed for healthy growth
Water	For photosynthesis
Space	So roots can absorb enough water and leaves can absorb enough light
Minerals	For photosynthesis and to keep their cells rigid

D The following is a list of some adaptations of plants and animals.
Suggest how each one helps the plant or animal to survive.

a Covered in thorns ____

b Good eyesight ____

c Very fast ____

d Covered in thick fur ____

e Stores water in stem ____

What you need to remember

In order to survive, plants and animals compete for _________________ such as water and space. Animals also
compete for _________________ to reproduce, and for _________________. Plants compete for water and for
_________________ for photosynthesis and _________________ for healthy growth. To help them survive, plants
and animals have special characteristics called _________________.

B3.2 Adapting to change

A Animals have a number of ways to cope with cold winter temperatures.
Complete the sentences below using the words from the box.

thicker fur　　**migration**　　**hibernation**　　**warm**　　**birds**

Animals like bears find somewhere _____________ to sleep through the winter. This is called

_____________. Animals like _____________ travel somewhere warmer in winter. This is

called _____________. Animals like sheep keep warm in winter by growing _____________

_____________.

B Some changes are so severe that only the best-adapted organisms can survive. Others have to leave the area or die.
Tick the sentences that could lead to this occurring.

W A new food source arrives in the habitat.　☐　　**Y** A new species which competes for food arrives in the habitat.　☐

X Fire destroys the habitat.　☐　　**Z** Food supply reduced by disease.　☐

C A population of rabbits lives in a field. Foxes are predators of rabbits.

Study the graph opposite.

a Label the lines to show which represents the population of foxes and which represents the population of rabbits.

b Tick the boxes next to all the correct statements.

W As the number of rabbits goes up, the number of foxes goes down.　☐

X When the number of rabbits goes down, the amount of grass in the field decreases.　☐

Y As fox numbers increase, the population of rabbits decreases.　☐

Z As the number of rabbits increases, so does the number of foxes.　☐

What you need to remember

Plants and animals have to cope with changes in their _____________. For example, in winter, trees lose their

_____________ and sheep grow thicker _____________. Sudden changes such as fire or disease mean that only the

best _____________ organisms survive and reproduce.

When a predator has only one main food source, there is an _____________ between the predator and prey

populations. A change in the population of one directly causes a change in the other. For example, if the prey

population increases, the predator population _____________ as they have more _____________, meaning more stay

alive to reproduce.

B3.3 Variation

A Complete the following sentences using the words in the box below.

variation	offspring	characteristics	species	identical

Organisms within the same _________________ have very similar _________________. For example, it is very hard to tell the difference between two blackbirds. These organisms can reproduce to produce fertile _________________. However no two organisms are _________________. The differences in characteristics within a species are called _________________.

B These leaves have all been taken from the same tree.

Give **two** ways the leaves vary.

1 ____

2 ____

C Variation between individuals may be due to inherited variation, environmental variation, or both.

Draw a line to match each human variation to its cause.

Variation

eye colour

height

accent

pierced ears

blood group

skin colour

Cause

inherited

environment

a combination of both

B3.4 Continuous and discontinuous

A Identify the **two** types of variation shown within a species. Tick the correct boxes.

Discontinuous variation ☐

Developed variation ☐

Climate variation ☐

Continuous variation ☐

B Which type of variation is being described?

a A characteristic that can take any value within a range. _______________________

b A characteristic that can only result in certain values. _______________________

C Choose the chart type that can be used to represent the following types of variation.

histogram	bar chart	tally chart	pictogram

a Continuous variation _______________________

b Discontinuous variation _______________________

D Sort the following characteristics into those that show continuous variation and those showing discontinuous variation.

leg length	blood group	leaf surface area
flower colour	fish mass	number of spots

Continuous variation	Discontinuous variation

What you need to remember

Characteristics that can only result in certain values show _______________ variation; for example, gender.

This type of variation should be plotted on a _______________________ . Characteristics that can

result in any value within a range show _______________ variation; for example, height. This type of variation

should be plotted on a _______________ .

B3.5 Inheritance

A Complete the following sentences about DNA, genes, and chromosomes using the words in the box below.

> **DNA**   **genes**   **chromosomes**

Your genetic information is stored on a long molecule called _________________ which is packaged into long

strands called _________________. Small sections of these molecules are called _________________ and contain

the information to produce a characteristic.

B Match each label on the diagram to the correct word below.

Write **S**, **T**, **U**, or **V** beside each word.

gene ☐

chromosome ☐

cell ☐

nucleus ☐

C You inherit characteristics from your parents through genetic material.
Circle the correct **bold** words in the sentences below to describe how genetic material is inherited.

Inside the nucleus of your cells, DNA is arranged into long strands called **genes / chromosomes**. You inherit

half / all your chromosomes from your mother and **half / all** your chromosomes from your father. Chromosomes

from your mother are carried in **an egg / a sperm** cell. Chromosomes from your father are carried in **an egg /

a sperm** cell. During **variation / fertilisation**, the nuclei of the two cells join, producing an embryo with a full

set of **23 / 46** chromosomes.

What you need to remember

You inherit characteristics from your parents through genetic material found in the _________________ of your

cells. Genetic material is made up of the chemical _________________ – this contains all the information needed

to make an organism. In the nucleus, this chemical is organised into long strands called _________________.

Each strand is divided into sections called _________________. Each section contains the information needed

to produce a characteristic.

B3.6 Natural selection

A Which of the following statements describes evolution? Tick **one** box.

Differences in characteristics within a species. ☐

When no more individuals of a species are left anywhere in the world. ☐

Changes in species over time. ☐

B The peppered moth is often used to study evolution. Circle the correct **bold** words in the sentences below to describe how this species has evolved.

There are two types of peppered moth: pale moths and dark moths. Before the industrial revolution there were more **pale / dark** moths as these were **camouflaged / highlighted** against the pale tree bark. This is because **pale / dark** moths were seen and eaten. More **pale / dark** moths survived and reproduced, **increasing / decreasing** the number of pale moths in the population.

After the industrial revolution, many trees were covered in soot. The **pale / dark** moths were now more camouflaged, so they survived and reproduced, and more **pale / dark** moths were eaten. This resulted in the number of pale moths **increasing / decreasing** and the number of dark moths **increasing / decreasing**.

Most moths in the population were then **pale / dark**, as these survived and reproduced.

C The statements below can be reordered to describe how a species evolves through the process of natural selection. Read the statements and write down the order of statements you think will give the best description.

Correct order ☐ ☐ ☐ ☐ ☐ ☐

1 Process is repeated over many generations.
2 Genes which code for advantageous characteristics are passed on to offspring.
3 New species can evolve where all organisms have the adaptations.
4 Organisms in a species show variation.
5 More organisms within the species have the advantageous characteristic.
6 The organisms with the characteristics that are best suited to the environment survive and reproduce. Less well-adapted organisms die.

D What two types of physical evidence do we have for evolution?

1 _________________ 2 _________________

What you need to remember

All organisms living today have _________________ from a common ancestor. This process has taken _________________ of years and has occurred as a result of _________________ _________________.

The organisms most adapted to their environment _________________ and reproduce, passing on the _________________ which code for these characteristics to their offspring. The remains of organisms that lived millions of years ago, called _________________, provide evidence for evolution.

B3.7 Extinction

A Match the definitions with the words in the box below.

biodiversity	extinct	endangered

The range of organisms living in an area: ________________.

Species of plants and animals that only have a small population in the world: ________________.

No individuals of a species living anywhere in the world: ________________.

B Which of the following techniques do scientists use to try to prevent a species becoming extinct?
Tick as many answers as you need.

Gene banks ☐

Poaching ☐

Captive breeding ☐

Habitat removal ☐

C Draw a line to match each factor with how it could lead to the extinction of a species.

Outbreak of disease	Whole population eaten before they have the chance to reproduce successfully
Prolonged drought	Lack of food and/or water causing death of whole population
Introduction of new predators	Whole population killed by a microorganism
Introduction of new competitors	Loss of food source or shelter leads to death of whole population
Deforestation	Lack of water leads to death of whole population

D Circle the correct **bold** words in the sentences below about gene banks.

Gene banks store **genetic / live** samples of different species.

They are normally stored at very **high / low** temperatures.

They can be used for research and to create new **individuals / species**.

What you need to remember

The range of organisms living in area is called ________________. Destruction of ________________ and

outbreaks of ________________ can cause a reduction in biodiversity and can lead to a species becoming

________________. This is where there are no individuals of a species living anywhere in the world. When only a

small number of a species exist, the species is said to be ________________. One way scientists try to ensure plant

species survive is through storing genetic material in a ________________ ________________.

Pinchpoint question

Answer the question below, then do the follow-up activity **with the same letter** as the answer you picked.

The graph shows the number of caribou (prey) in a herd found in Quebec and a population of wolves (predator) found in the same area.

Read through the statements below and choose which one best describes the trends shown in the graph.

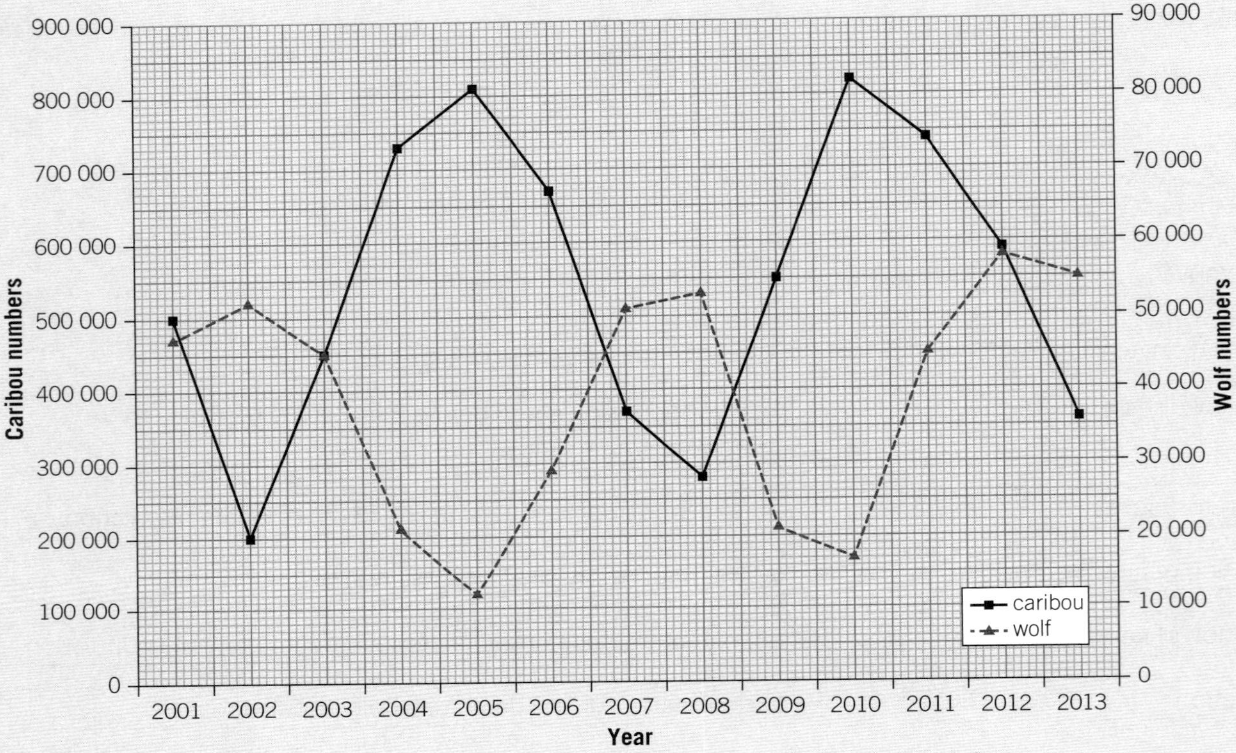

A Wolves and caribou alternate having the highest population; this causes fluctuations on the graph.

B When the predator population increases, the prey population also increases.

C When the caribou population decreases, the wolf population increases some time later.

D When the prey population increases, the predator population increases some time later.

Follow-up activities

A **a** Name the year when the population of wolves was at its highest. _____________

  b Write down the population of wolves in this year. _____________________

  c Write down the population of caribou in this year. _____________________

  Hint: Look carefully at the axis labels. You can draw lines on the graph to make sure your data readings are accurate. See B2 3.2 Adapting to change or B1 1.4 Analysing data for help.

B **a** Look at the graph and circle the correct endings to these sentences.

Between 2002 and 2005, the caribou population **increases / decreases**.

Between 2002 and 2005, the wolf population **increases / decreases**.

Between 2005 and 2008, the caribou population **increases / decreases**.

Between 2005 and 2008, the wolf population **increases / decreases**.

b Complete the sentences to describe the trends in predator–prey relationships.

i As the prey population increases, ____

ii As the prey population decreases, ____

Hint: Look at each section of the graph in turn. Try highlighting the data line to make it easier to see the trend. See B2 3.2 Adapting to change or B1 1.4 Analysing data for help.

C The statements below can be reordered to describe the trends shown in the graph. Read the statements and write down the order of statements you think will give the best description.

Correct order:

1 The wolves survive longer and reproduce more, increasing their population.

2 Fewer wolves are present to feed on caribou, so the caribou population increases.

3 Eventually there is not enough food for all the predators, so their numbers decrease.

4 When the caribou population increases, the wolves have more to eat.

5 More predators are now present, so they eat more prey; prey numbers fall.

Hint: Predators need to have prey available to eat. Without this, they will die. See B2 3.2 Adapting to change for help.

D Imagine that, over time, caribou evolve to run faster. Suggest what could happen to the population of wolves, and the population of caribou.

____

____

____

____

____

____

Hint: What would happen if the wolves didn't evolve as well? What would happen if they did? See B2 3.2 Adapting to change for help.

⧓ Pinchpoint review

Now look back at the question – do you think you chose the right letter?
Turn to the Answers page to find out.

B2 Revision questions

1 Food chains show the transfer of energy between organisms.

  a Organise the following organisms into a food chain. One organism has been done for you. *(3 marks)*

greenfly tomatoplant sparrow ladybird hawk

  ______________ → ______________ → ladybird →

  ______________ → ______________

  b Compete the following sentence. *(1 mark)*

  A food chain shows the transfer of ______________ between organisms.

2 Choose **one** drug from the box below to answer each of the following questions. (You can use each drug more than once.)

caffeine	**paracetamol**	**tobacco**
ethanol (alcohol)		**antibiotic**

  a An example of a recreational drug. *(1 mark)*

  ____

  b An example of a medicinal drug. *(1 mark)*

  ____

  c A drug that can cause lung cancer. *(1 mark)*

  ____

  d A drug that slows down body reactions. *(1 mark)*

  ____

3 Complete the following sentences using words from the box below. *(4 marks)*

genes	**nucleus**	**characteristic**
	DNA	**chromosomes**

  Your genetic information is stored in the chemical ______________ found in the ______________ of your cells. This chemical is arranged into long strands called ______________. Small sections of these structures are called ______________. Each one contains the information needed to produce a ______________.

4 A polar bear has special features that help it live in the Arctic.

  a These features are called ______________. *(1 mark)*

  b Give two reasons why polar bears have thick white fur. *(2 marks)*

  ____

  ____

  ____

5 **Figure 1** shows the human digestive system.

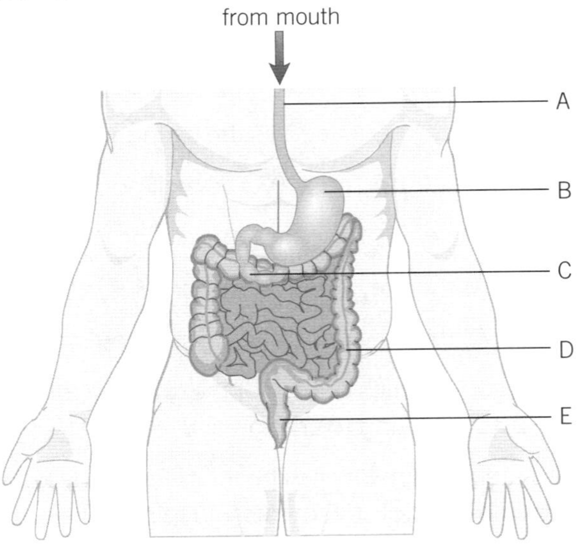

Figure 1

  a Identify which label on the diagram is pointing to the following structures.

  Stomach ______________ *(1 mark)*

  Gullet ______________ *(1 mark)*

  Large intestine ______________ *(1 mark)*

  b Describe what happens during digestion. *(2 marks)*

  ____

  ____

  ____

6 To remain healthy, you should eat a balanced diet.

  a Name two nutrients that form part of a balanced diet. *(2 marks)*

  ____

  ____

  b Name one health problem a person may have if they eat too much fat. *(1 mark)*

  ____

  c Name one health problem a person may have if they do not eat enough food. *(1 mark)*

  ____

7 Plants make their own food by the process of photosynthesis.

  a Give **two** products of photosynthesis. *(2 marks)*

  ____

  ____

  b Name where photosynthesis occurs in a cell. *(1 mark)*

  ____

c Plants store glucose in their leaves as starch. **Figures 2** and **3** show two of the main stages used when testing a leaf for starch.

Describe the purpose of each step.

i *(2 marks)*

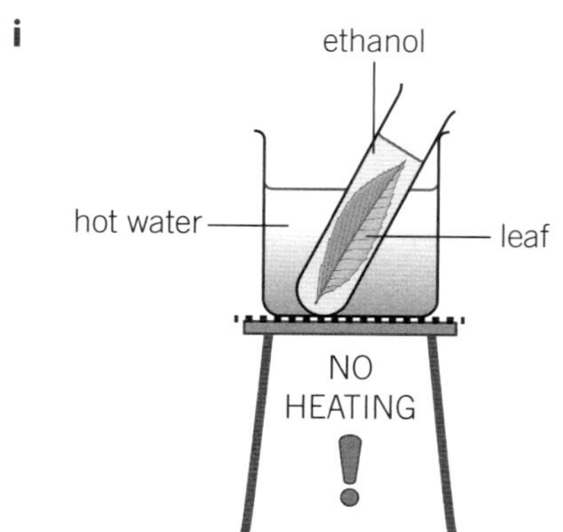

Figure 2

____

____

____

ii *(2 marks)*

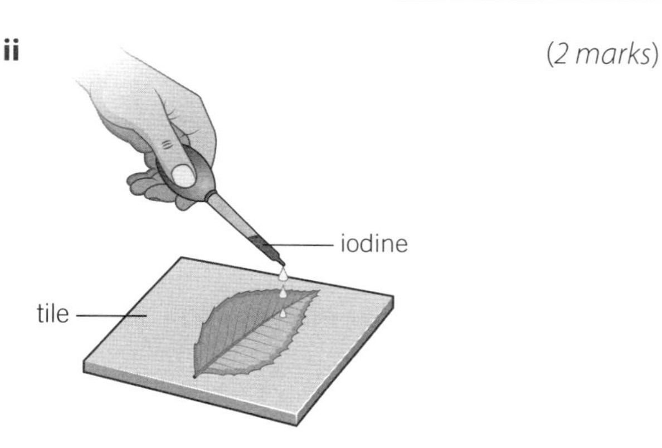

Figure 3

____

____

____

d Explain why the Bunsen burner must be turned off before the step shown in **Figure 2.** *(1 mark)*

____

8 Organisms have special adaptations to enable them to survive in their environment.

a Describe how the following adaptations enable a cactus to survive in a desert. *(3 marks)*

i waxy layer _____________________

____

ii spikes_____________________________

____

iii widespread roots _____________________

____

b Organisms have developed adaptations to cope with seasonal changes. Describe **one** way a plant and **one** way an animal copes with the change from summer to winter. *(2 marks)*

i Plant adaptation: _____________________

____

ii Animal adaptation: _____________________

____

9 **Figure 4** is a food web which shows the feeding relationships between organisms living in a grass field.

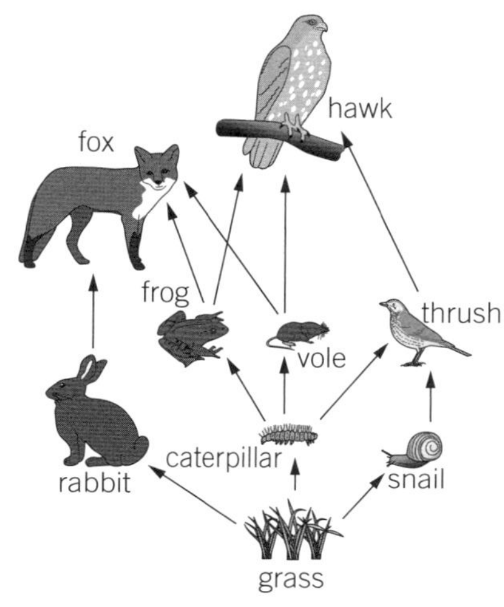

Figure 4

a Describe what would happen to the rabbit population if the fox population was removed. *(1 mark)*

____

b Describe what would happen to the vole population if the caterpillar population was removed. *(1 mark)*

____

c An insecticide was sprayed to kill off the snail population. Chemicals from the insecticide remained in the body of the snails. Describe how this can result in the death of hawks. *(3 marks)*

____

____

____

____

10 ⚗ ⚗ A group of students were investigating the types of nutrient present in different ready meals.

a Draw a line to match each nutrient to its role in the body. *(2 marks)*

Protein		Main source of energy
Lipids		To provide a store of energy and insulation, and protect organs
Carbohydrates		To repair body tissues and make new cells

b The ready meals also contained fibre. Explain the importance of fibre in a person's diet. *(2 marks)*

c One of the ready meals stated that it contained 2520 kJ of energy per pack. An average adult requires 8400 kJ of energy per day.
Calculate the percentage of an adult's daily energy requirement contained within this pack. *(2 marks)*

________ kJ

d Describe how the students could prove that sugar was present in one of the ready meals *(3 marks)*

11 ⚗ ⚗ Digestion begins in your mouth when food is chewed and mixed with saliva.

a i Name the enzyme present in saliva. *(1 mark)*

ii Describe the role of this enzyme in digestion. *(2 marks)*

b Explain why enzymes are called biological catalysts. *(2 marks)*

c Bacteria that live in your large intestine play an important role in digestion. Describe the role these bacteria play. *(2 marks)*

12 ⚗ ⚗ You get energy from the food you eat. This energy is transferred to your cells by respiration.

a i Complete the word equation for aerobic respiration. *(2 marks)*

glucose + ________ →

carbon dioxide + ________

ii Write down where in the cell respiration takes place. *(1 mark)*

b Write down **two** differences between aerobic and anaerobic respiration in humans. *(2 marks)*

c i Write down what type of respiration is being represented by the following word equation:

glucose → ethanol + carbon dioxide (+ energy) *(1 mark)*

ii Name an organism that performs this type of respiration. *(1 mark)*

13 A group of Year 8 students studied variation within their class.

a Suggest one characteristic that they could investigate which shows:

i environmental variation *(1 mark)*

____

ii inherited variation *(1 mark)*

____

b The students also collected data on variation within holly leaves. Their data are shown in **Table 1**.

Table 1

Number of spikes	8	9	10	11	12
Frequency	8	10	15	12	14

i Plot the data using a bar chart on the graph paper below. *(4 marks)*

ii Give the most common number of spikes found on a holly leaf. *(1 mark)*

____

c **i** Name the type of variation shown by this characteristic. *(1 mark)*

____

ii Suggest how light could have caused this type of variation. *(3 marks)*

____

____

____

____

14 Plants and animals compete for resources to survive. For example, plants compete for light.

a Describe one other resource that plants compete for. *(2 marks)*

____

____

b Explain why animals do not compete for light. *(2 marks)*

____

____

____

c A new predator is introduced into an area. Describe how its prey species could change over a long period of time to survive a new predator. *(6 marks)*

____

____

____

____

____

____

____

____

B2 Checklist

Revision question number	Outcome	Topic reference	☹	😐	☺
1a, b	State the definition of a food chain.	B2 2.7			
2a	Name some recreational and medicinal drugs.	B2 1.6			
2b	Name an effect of tobacco smoke on health.	B2 1.8			
2c	Name an effect of alcohol on health or behaviour.	B2 1.7			
2d	State whether or not alcohol affects pregnancy. State whether or not tobacco affects the fetus.	B2 1.7 B2 1.8			
3	State what is meant by a gene.	B2 3.5			
4a	Name the parts of the digestive system.	B2 1.4			
4b	State what is meant by digestion.	B2 1.4			
5a	State what is meant by the term adaptation.	B2 3.1			
5b	Describe how organisms are adapted to their environments.	B2 3.1			
6a	Name some nutrients in a diet.	B2 1.1			
6b, c	State potential problems for someone with an unhealthy diet.	B2 1.3			
7a	State the products of photosynthesis.	B2 2.1			
7b	State where photosynthesis occurs in a plant.	B2 2.1			
7c, d	Carry out an experiment to test for the presence of starch.	B2 2.1			
8a	Describe how organisms are adapted to their environment.	B2 3.1			
8b	Describe how organisms adapt to environmental changes.	B2 3.2			
9a, b	Describe the interdependence of organisms.	B2 2.8			
9c	Describe how toxic materials can accumulate in a food web.	B2 2.8			
10a, b	Explain the role of each nutrient in the body.	B2 1.1			
10c	Calculate energy requirements.	B2 1.3			
10d	Describe how to test foods for the presence of sugars.	B2 1.2			
11a, b	Describe the role of enzymes in digestion.	B2 1.5			
11c	Describe the role of bacteria in digestion.	B2 1.5			
12a	State the word equation for aerobic respiration.	B2 2.5			
12b, c	Describe the differences between aerobic and anaerobic respiration.	B2 2.5			
13a	Describe the difference between environmental and inherited variation.	B2 3.3			
13b	Represent variation within a species using graphs.	B2 3.4			
13ci	Describe the difference between continuous and discontinuous variation.	B2 3.4			
13cii	Describe an example of environmental variation.	B2 3.3			
14a, b	Describe some resources that plants and animals compete for.	B2 3.1			
14c	Describe the process of natural selection.	B2 3.6			

C1.1 Metals and non-metals

A The diagram shows a Periodic Table, without the names of the elements.

Label the diagram using the words provided.

| metals | non-metals |

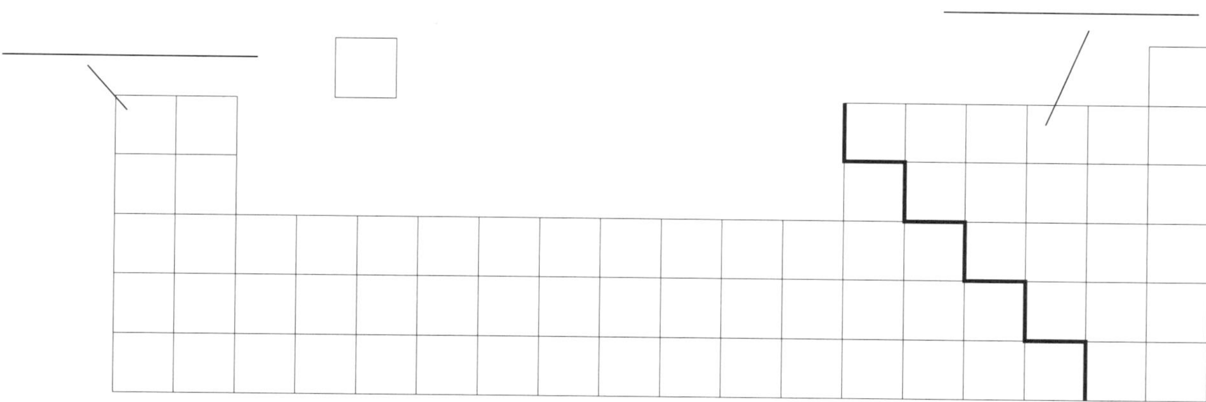

B The elements can be classified into metals and non-metals.
Highlight or underline each statement below that describes a property of a typical **non-metal.**

It is shiny.

It is a good conductor of electricity.

It is brittle.

It is sonorous, meaning it makes a
ringing sound when hit.

It is a gas at room temperature.

It is a poor conductor of heat.

It is ductile, meaning that you can pull it into wires.

It is malleable.

C The table shows the properties of four elements. Each element is represented by a letter.
The letters are not the same as the chemical symbols for the elements.

Element	State at room temperature	Density at room temperature (g/cm³)	Does it conduct electricity?
W	solid	19	yes
X	gas	1	no
Y	solid	7	yes
Z	solid	7	yes

Write down the letters of the elements in the table that are likely to be metals. _______________

What you need to remember

The elements are classified into metals and ____-__________. Metals are on the __________________ of the stepped

line in the Periodic Table, and non-metals are on the __________________ of the stepped line. Most metals have

__________________ melting and boiling points. They are __________________ conductors of heat and electricity.

Metals make a ringing sound when you hit them – in other words, they are __________________. The properties

listed so far are properties that you can observe and measure, so they are __________________ properties. Metals

and non-metals also have different chemical reactions, so their __________________ properties are different.

A The table shows the melting points of the elements in Group 1 of the Periodic Table.

Plot the melting point values on a bar chart, using these axes.

Element	Melting point (°C)
lithium	180
sodium	98
potassium	64
rubidium	39

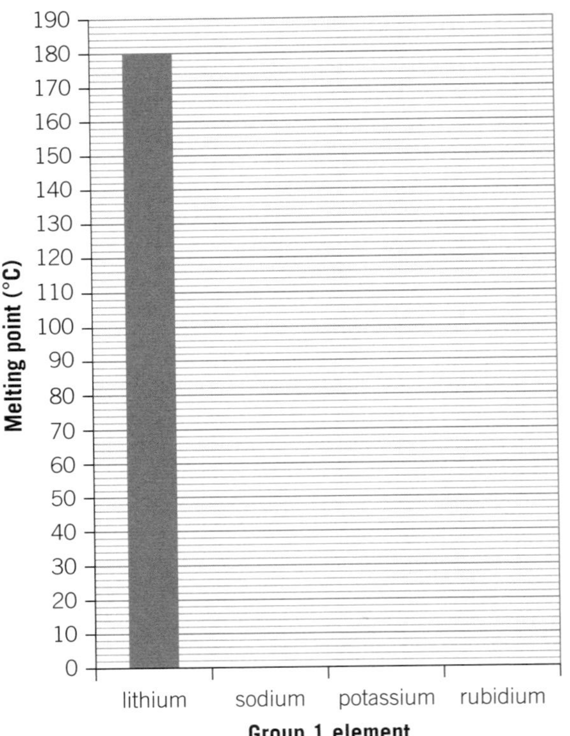

B Circle the correct **bold** words in the sentences below.
Use data from the table in activity **A** to help you, and the Periodic Table.

Lithium is at the **top / bottom** of Group 1, and rubidium is near

the **top / bottom** of the group. From top to bottom of the group,

the melting point **increases / decreases**. The element caesium

is **above / below** rubidium in Group 1. A sensible prediction for

the melting point of caesium is **29 / 49** °C.

C The tables shows the densities of some Group 3 and Group 4 elements. Boron is at the top of Group 3 of the Periodic Table, and carbon is at the top of Group 4.

Group 3 element	Density (g/cm³)
boron	2.3
aluminium	2.7
gallium	5.9
indium	7.3
thallium	11.8

Group 4 element	Density (g/cm³)
carbon	2.2
silicon	2.3
germanium	5.3
tin	7.3
lead	11.3

Draw a line from each sentence starter to **one** correct ending.
You can use each ending once, twice, or not at all.

| From top to bottom of Group 3, | | similar for Group 3 and Group 4. |

| From top to bottom of Group 4, | | density increases. |

| The pattern in density is | | different for Group 3 and Group 4. |

| From bottom to top of Group 4, | | density decreases. |

What you need to remember

In the Periodic Table, the vertical columns are called ________________ and the horizontal rows are called

________________. There are patterns in the properties of the elements down ________________ and across

________________. You can use patterns in the melting point of the elements in a ________________ to predict

the melting point of an element whose melting point you do not know.

C1.3 The elements of Group 1

A Mr Guthrie adds some universal indicator to a big container of water.
He adds a small piece of potassium, and there is a vigorous reaction.
Label the diagram using the words provided.

water and universal indicator	potassium	purple flame	bubbles of gas

B The sentences below are about the experiment in activity **A**, the reaction of potassium and water. There is one mistake in each sentence or word equation. Copy out each sentence, correcting its mistake.

a One of the products of the reaction is oxygen gas.

b The other product of the reaction is potassium oxide.

c potassium + water → potassium hydroxide + oxygen

d The universal indicator changes colour from green to purple, so the pH at the end could be 2.

e Rubidium is below potassium in Group 1, so the reaction of rubidium with water is less vigorous.

C The table shows the atomic radius of some elements in Group 1. The elements in the table are shown in the same order as they are in the Periodic Table. The greater the atomic radius, the bigger the atom.

Circle the correct **bold** words and phrases in the sentences below.

Element	Atomic radius (nm)
sodium	0.19
potassium	0.23
rubidium	0.25

Going down Group 1, the atomic radius **increases / decreases**. Lithium is at the top of Group 1, above sodium.

The pattern suggests that the atomic radius of lithium is **less than / more than** the atomic radius of sodium.

Caesium is below rubidium in Group 1. The pattern suggests that its atomic radius could be **0.21 / 0.26** nm.

What you need to remember

Group 1 contains the elements in the column on the _________________ of the Periodic Table. The Group 1 elements are metals. They _________________ electricity and have _________________ densities. The Group 1 elements react vigorously with water – in other words, they are very _________________. When a Group 1 element reacts with water, _________________ substances are made. These substances are _________________ gas and a metal _________________.

A Mrs Hull is using bromine to do an experiment. The bottle of bromine has two hazard symbols.
Draw a line to match each hazard symbol to its meaning, a risk from the hazard, and how to control the risk.

Hazard symbol	Meaning	Risk from this hazard	How to control the risk
	toxic	burns eyes	use in a fume cupboard, so that any fumes from the liquid do not go into the room
	corrosive	difficulty breathing	wear safety goggles

B The table shows some data for Group 7. Fluorine is at the top of the group.

Element	State at room temperature	Colour at room temperature
fluorine	gas	pale yellow
chlorine	gas	green
bromine	liquid	dark red
iodine	solid	grey-black

Circle the correct **bold** words in the sentences below. Use data from the table to help you.

The colours of the Group 7 elements get **lighter / darker** going down the group. The elements at the **top /**

bottom of the group are in the gas state at room temperature. Astatine is at the bottom of Group 7, below

fluorine / iodine. This means that it is sensible to predict that astatine is in the **solid / liquid / gas** state at room

temperature and that its colour is **dark / light**.

C Mrs Hull adds chlorine solution to sodium bromide solution. Her observations are in the table.

Solution	Appearance
chlorine solution (before reaction)	pale green
sodium bromide solution (before reaction)	colourless
mixture after reaction	yellow-orange

Tick the statements about the experiment that are true.

1 The products of the reaction are sodium chloride and bromine ☐

2 The reaction is a displacement reaction. ☐

3 Chlorine is less reactive than bromine. ☐

4 Chlorine displaces bromine from sodium bromide. ☐

What you need to remember

Group 7 contains the elements in the column that is second from the _______________ of the Periodic Table.

The elements in Group 7 are also called the _______________. They are non-_______________. The Group 7

elements take part in displacement _______________. In one of these, chlorine reacts with sodium bromide to

make sodium _______________ and bromine. In this reaction, chlorine _______________ bromine from one

of its compounds.

C1.5 The elements of Group 0

A Choose the **three** correct statements that describe the elements in Group 0.

The elements in Group 0… ___________________ ___________________ ___________________

A	are unreactive.	**D**	are called the halogens.
B	are in the gas state at room temperature.	**E**	react vigorously with water.
C	are in the solid state at room temperature.	**F**	are called the noble gases.

B The bar chart shows the boiling points of the Group 0 elements. The boiling point of krypton is missing.

Tick the statements below that are true.

1 All the noble gases have boiling points above 0 °C. ☐

2 The boiling point of argon is nearer to 0 °C than the boiling point of neon. ☐

3 The boiling point of argon is lower than the boiling point of neon. ☐

4 From helium at the top of Group 0 to xenon at the bottom, boiling point increases. ☐

5 Boiling point decreases from top to bottom of Group 0. ☐

6 From the pattern shown, a sensible prediction for the boiling point of krypton is −152 °C. ☐

C The table gives the density of each Group 0 element, and air.

a Write the chemical symbols of the elements in order of **decreasing** density, highest density first.

____

____

Element	Chemical symbol	Density in kg/m³ at 0 °C
helium	He	0.2
neon	Ne	0.9
argon	Ar	1.8
krypton	Kr	3.7
xenon	Xe	5.9
air	–	1.3

b Each Group 0 element is used to fill a separate balloon.

Predict which balloons fall to the ground when dropped in air.

____

Hint: Balloons filled with a gas that is more dense than air fall, and balloons filled with a gas that is less dense than air rise.

What you need to remember

Group 0 contains the elements in the column on the ___________________ of the Periodic Table. The elements in

Group 0 are also called the ___________________ gases. They are non-___________________. Most Group 0 elements

do not take part in chemical reactions – in other words, they are ___________________.

Pinchpoint question

Answer the question below, then do the follow-up activity **with the same letter** as the answer you picked.

Look at these elements in Group 1 and Group 7 of the Periodic Table.

Li lithium																	F fluorine
Na sodium																	Cl chlorine
K potassium																	Br bromine
Rb rubidium																	I iodine
Cs caesium																	

Every Group 1 element can react with every Group 7 element.

Which pair of elements reacts most vigorously, and why?

A   Sodium and chlorine because they are both very reactive elements.

B   Potassium and chlorine because potassium is more reactive than the elements above it and chlorine is more reactive than the elements below it.

C   Caesium and fluorine because the Group 1 elements are more reactive from top to bottom but the Group 7 elements get less reactive from top to bottom.

D   Caesium and iodine because the Group 1 elements get more reactive from top to bottom and the same is true for all other groups.

Follow-up activities

A   In Group 1, elements get **more** reactive from top to bottom. In Group 7, elements get **less** reactive from top to bottom.

Circle the correct **bold** words in the sentences below.

Caesium is at the **top / bottom** of **Group 1 / Group 7**. This means that caesium is the **most / least** reactive element in this group. Lithium is at the **top / bottom** of **Group 1 / Group 7**. This means that lithium is the **most / least** reactive element in this group. Fluorine is at the **top / bottom** of **Group 1 / Group 7**. This means that fluorine is the **most / least** reactive element in this group. Iodine is near the **top / bottom** of **Group 1 / Group 7**.

This means that iodine is **more / less** reactive than fluorine, chlorine, and bromine.

Hint: In Group 1, the most reactive element is at the bottom of the group, and in Group 7 the most reactive element is at the top of the group. Which is the most reactive element in Group 1? For help see C2 1.3 The elements of Group 1 and C2 1.4 The elements of Group 7.

B   In Group 1, elements get **more** reactive from top to bottom. In Group 7, elements get **less** reactive from top to bottom.

Tick the statements below that are true. Consider only the elements shown in the partial Periodic Table above.

1   The most vigorous reaction is between caesium and iodine. ☐
2   The least vigorous reaction is between lithium and iodine. ☐
3   The reaction of lithium and fluorine is more vigorous than the reaction of sodium and fluorine. ☐
4   The reaction of caesium and bromine is more vigorous than the reaction of caesium and iodine. ☐

Hint: The most vigorous reaction is between the most reactive element in Group 1 and the most reactive element in Group 7. Which is the most reactive element in Group 1? For help see C2 1.3 The elements of Group 1 and C2 1.4 The elements of Group 7.

C Look at the lists of elements in Group 2 and Group 7. The elements are in the same order as in the Periodic Table.

Group 2

beryllium
magnesium
calcium
strontium
barium

Group 7

fluorine
chlorine
bromine
iodine

In Group 2 the elements get **more** reactive from top to bottom, just like Group 1.

In Group 7 the elements get **less** reactive from top to bottom.

List the pairs of elements below in decreasing order of how vigorously they react, **most reactive first**.

Correct order ☐ ☐ ☐ ☐

1   beryllium and iodine

2   barium and fluorine

3   barium and chlorine

4   magnesium and iodine

Hint: The most vigorous reaction is between the most reactive elements in each group. Which is the most reactive element in Group 7? For help see C2 1.7 The elements of Group 7.

D The patterns in reactivity are different for different groups of the Periodic Table:

In Groups 1 and 2 the elements get **more** reactive from top to bottom.
In Group 7 the elements get **less** reactive from top to bottom.

Draw a line to match each element name to its position in its group and to its relative reactivity.

Element	Position and group	Relative reactivity
Lithium	is at the top of Group 1	so it is the least reactive element in its group.
Fluorine	is at the bottom of Group 1	
Caesium	is at the top of Group 7	so it is the most reactive element in its group.
Iodine	is near the bottom of Group 7	

Hint: In Group 1, the most reactive element is at the bottom of the Group, and in Group 7 the most reactive element is at the top of the Group. For help see C2 1.3 The elements of Group 1 and C2 1.4 The elements of Group 7.

 **Pinchpoint review**

Now look back at the question – do you think you chose the right letter?
Turn to the Answers page to find out.

C2.1 Mixtures

A Highlight or underline each statement below that describes a mixture.

1 The different substances in a mixture are not joined together.

2 Each substance in a mixture has its own melting point.

3 You can change the amounts of the different substances in a mixture.

4 It is always very difficult to separate the different substances in a mixture.

B Draw a line from each mixture to show how to separate it.

sand and water	with a magnet
sand and steel nails	with filter paper in a filter funnel
flour and marbles	with a sieve

C A pure substance has a sharp melting point.
An impure substance melts over a range of temperatures.

A student measures the melting temperatures of four substances: **W**, **X**, **Y**, and **Z**. Her results are in the table.

Substance	Temperature the substance started to melt at (°C)	Temperature that the substance finished melting at (°C)
W	11	13
X	51	51
Y	37	53
Z	79	79

Write down the letters of the **two** pure substances in the table. __________________ __________________

D The diagrams show some pure substances and mixtures, all in the gas state.
Each circle represents one atom. Different-coloured circles represent atoms of different elements.

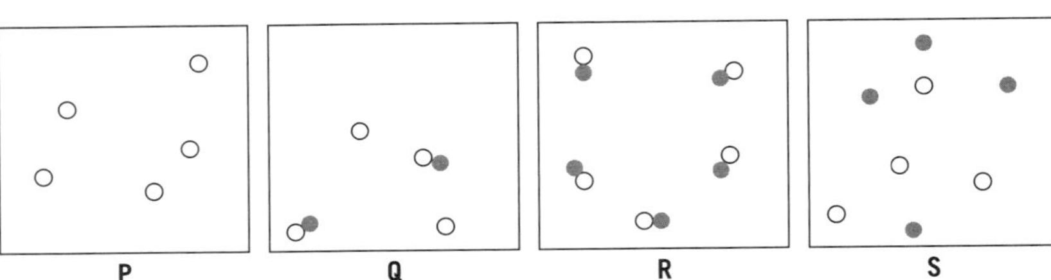

Circle the correct **bold** words in the sentences below.

In box P all the atoms are **the same / different**. This means that box P contains **one element / a compound**.
Box S has atoms of **one / two** elements. The atoms of the different elements are not joined together, so
box S is a **mixture / compound** of **elements / compounds**. Box Q is a mixture of an element and
another element / a compound.

What you need to remember

If a substance does not have other substances mixed with it, it is __________________ and has a __________________

melting point. If a substance has other substances mixed with it, it is __________________ and melts over

a range of __________________. A mixture contains __________________ than one substance. Its different

substances are __________________ joined together. You can change the amounts of the different substances in a

__________________, and the substances in a mixture keep their own __________________.

A Brett adds some salt to water, and stirs the mixture.

The sentences below are about Brett's experiment. There is one mistake in each sentence.

Copy out each sentence, correcting its mistake.

a A mixture of salt and water, when all the salt is dissolved, is called a solvent.

____

b In Brett's experiment, salt is the solvent.

____

c In Brett's experiment, water is the solute.

____

B The diagram shows some sugar dissolved in water. Label the diagram using the labels provided.

| **sugar particle** | **water particle** |

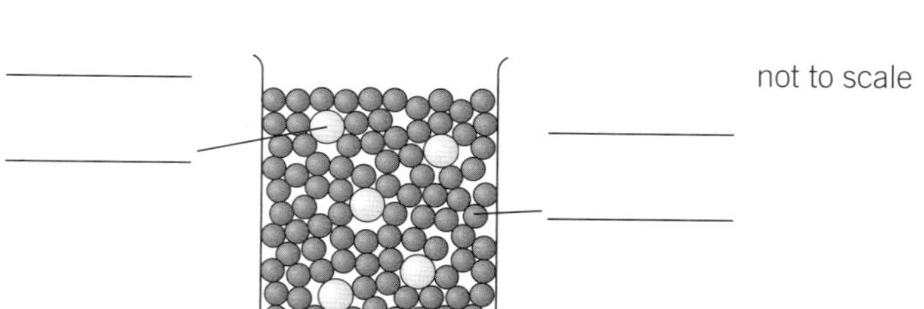

not to scale

C Circle the correct **bold** words and phrases in the sentences below.

In solid sugar, sugar particles are **far apart / close together**. They are arranged **in a regular pattern / randomly**.

When you add solid sugar to water, the sugar particles **get closer together / separate from each other**.

Water particles **surround / move away from** each sugar particle, making a solution. In a solution, the particles

are arranged **randomly / in a pattern** and they **move around / cannot move**.

D Brooke plans to add different solutes to water to make four solutions: **W**, **X**, **Y**, and **Z**.

She measures the mass of the water and the mass of the solute she uses to make each solution.

She also measures the mass of each solution.

Complete the table by writing **one** number in each empty box.

Solution	Mass of water (g)	Mass of solute (g)	Mass of solution (g)
W	100	9	
X	90	3	
Y	100		110
Z	85		90

What you need to remember

When you mix sugar with water, the sugar dissolves to make a _______________. Sugar is the

_______________ and water is the _______________. In a solution, several solvent particles surround each

_______________ particle. The particles _______________ around. In a solution, you _______________ see

two separate substances, and all parts of the mixture are the _______________. In a solution, the solvent is in the

_______________ state and the substance that dissolves in it can be in the solid or _______________ state.

C2.3 Solubility

A Draw a line to match each word or phrase to its definition.

Solute	A solution in which no more solute can dissolve.
Solvent	The solid or gas that dissolves in a liquid.
Soluble	If a substance dissolves in a solvent, the substance is ______________ in that solvent.
Saturated solution	The liquid in which a solid or gas can dissolve.
Solubility	The mass of solute that dissolves in 100 g of water to make a saturated solution.

B The graph shows the solubility of potassium nitrate at different temperatures.

a Draw a line of best fit through the points.

b Circle the correct **bold** words and numbers in the sentences below.

At 20 °C, the solubility of potassium nitrate is

33 / 65 g per 100 g of water. At **40 /60** °C, its

solubility is 106 g per 100 g of water. At 100 °C,

the solubility of potassium nitrate is **167 / 240** g

per 100 g of water.

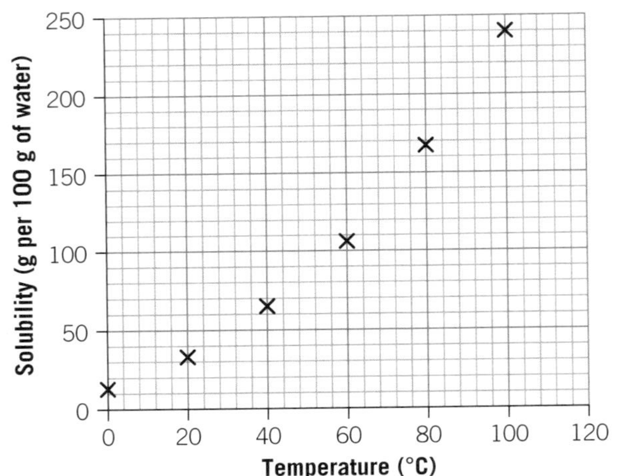

Overall, as temperature increases, the solubility of

potassium nitrate **decreases / increases**. This is

true for most solutes, but not all.

C The statements below can be reordered to describe an experiment to investigate the solubility of a solute at different temperatures. Read the statements and write down the order of statements you think will give the best method.

Correct order ☐ ☐ ☐ ☐ ☐

1 Heat up some water in a kettle, and repeat the whole experiment four more times at different temperatures.

2 Continue to add more and more solute to the water, with stirring, until no more dissolves.

3 At room temperature, weigh out 100 g of water in a beaker.

4 Use a spatula to add some solute to the water, and stir with a stirring rod.

5 Record the final mass of the solution.

What you need to remember

A solution in which no more solute can dissolve is called a ______________ solution. The mass of solute that

dissolves in 100 g of water to make a saturated solution is the ______________ of the solute. If a substance

dissolves in a solvent, scientists say that the substance is ______________. The greater the mass of solute that

can dissolve, the more ______________ the solute.

C2.4 Filtration

A Label the diagram using the labels in the box below.

| filter paper cone | filter funnel | conical flask | residue | filtrate | clamp |

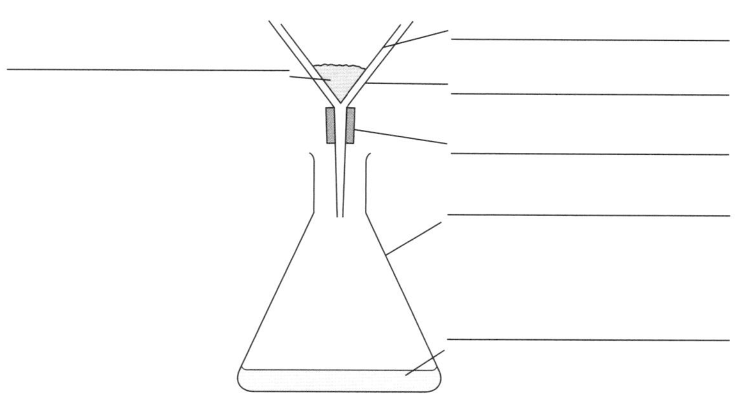

B Complete the table by writing the names of the residue and filtrate for each mixture.

Mixture	Residue	Filtrate
Sand and water		
Pieces of dirt from oil in a car engine		
Coffee solution from ground-up coffee beans		

C Ben filters a mixture of sand and water. This is what he does.

1 Put a filter funnel in a test tube.

2 Make a hole in the centre of the filter paper and fold it to make a cone.

3 Put the filter paper cone in the test tube.

4 Pour the mixture through the hole in the filter paper cone.

There is one mistake in each step. Use a coloured pen to correct each mistake.

D Match the sentence starters and endings below to make three correct statements.

Filter paper	smaller than the tiny holes, so they pass through the filter paper.
Water particles are	bigger than the tiny holes, so they stay in the filter paper cone.
Grains of sand are	has tiny holes in it.

What you need to remember

Filtration, also called ______________, separates a liquid from an ______________ solid. It can also separate a solution from a solid if the solid is not ______________. When you pour the mixture into filter paper, the ______________ or solution goes through tiny holes in the filter paper. The liquid or solution is the ______________. The solid remains in the filter paper cone. This is the ______________.

C2.5 Evaporation and distillation

A Complete the table to show whether you can obtain the substance from its mixture by distillation or evaporation.

Mixture	Substance wanted	You can obtain the substance by ....
salt solution	salt	
a mixture of ink and water	pure water	
copper chloride solution	copper chloride crystals	
salt solution	water	

B The statements below can be reordered to describe how to obtain big copper sulfate crystals from copper sulfate solution. Read the statements and write down the order that will give the best method.

Correct order ☐ ☐ ☐ ☐ ☐

1 Use a Bunsen burner to heat the beaker of water so that it boils.

2 Remove the evaporating basin and place it in a warm, dry place for a few days.

3 Pour the solution into an evaporating basin.

4 Continue to heat until you can see small crystals of copper sulfate around the edge of the evaporating basin.

5 Place the evaporating basin and its contents on top of a big beaker of water.

C Mr Pyle uses the apparatus below to obtain pure water from a mixture of ink and water. Label the diagram using the labels in the box below.

pure water	**inky water**	**condenser**	**Bunsen burner**
thermometer	**beaker**	**tap water in**	**tap water out**

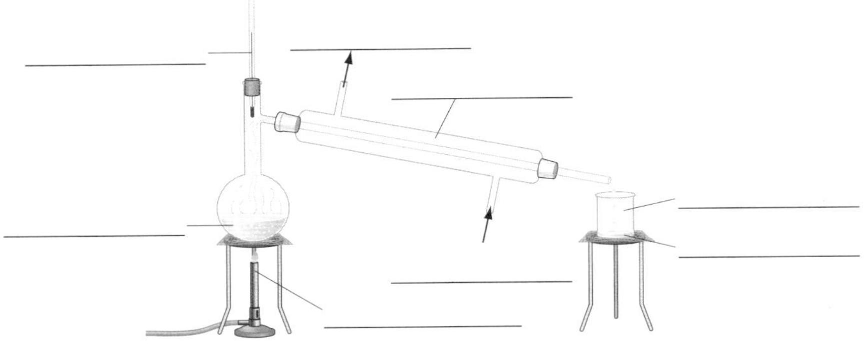

D Circle the correct **bold** words in the sentences below.

In distillation, the first step is to **heat / cool** the solution. Particles of **solute / solvent** leave the solution in the gas state. The solvent enters the condenser in the **gas / liquid** state. Here, the gas cools and **condenses / melts**. The solvent is now in the **solid / liquid** state. The solvent then flows down the condenser, and drips into the **beaker / condenser**.

What you need to remember

To obtain salt from salty water, you can use ________________. The water evaporates, and salt (in ________________ state) remains. Evaporation is used to obtain any ________________ from its solution.

To obtain water from salty water, you need to use ________________. Distillation uses evaporation and ________________ to obtain a solvent from its ________________.

C2.6 Chromatography

A Natalya sets up the apparatus below to find out about the dyes that are mixed together in ink. Tick the statements below that are true.

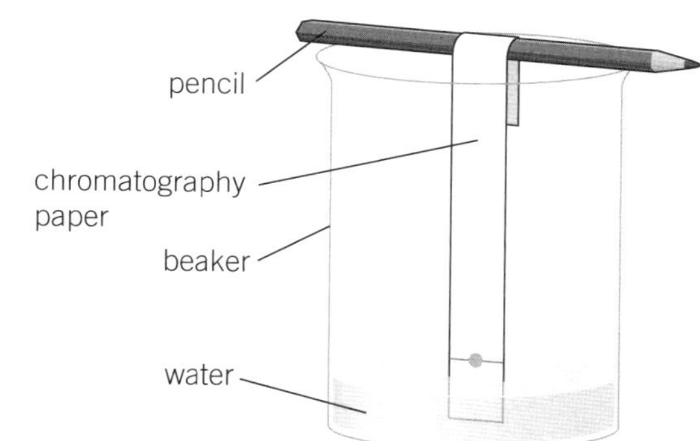

  1 Natalya must use pen to draw a line near the bottom of the chromatography paper. ☐

  2 The dyes that are mixed in the ink spot move up the paper. ☐

  3 The dyes in the ink spot move different distances. ☐

  4 The dyes in the ink spot are separated. ☐

B Here is Natalya's chromatogram.

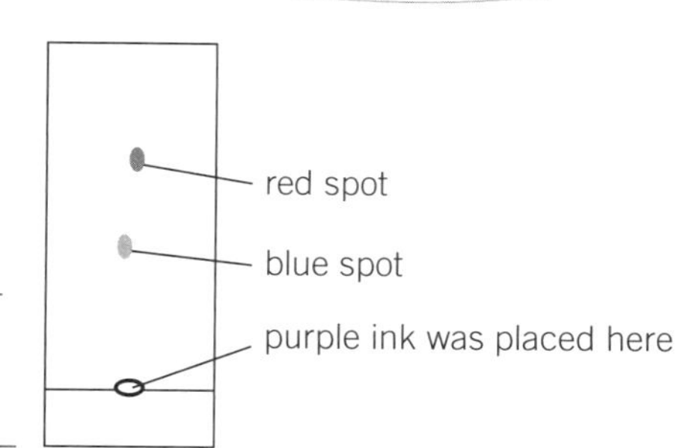

Name which colour ink each of the following statements is describing.

This colour ink is more soluble in water. _______________________

This colour ink sticks more strongly to the paper. _______________________

This colour ink is less soluble in water. _______________________

This colour ink is attracted less strongly to the paper _______________________

C Sam grinds up some leaves with a solvent, and puts a spot of pigment on a piece of chromatography paper. On the same paper, he also puts spots of three other pigments (colours) that he thinks might be in the leaves. These are his reference spots.

Here is Sam's chromatogram.

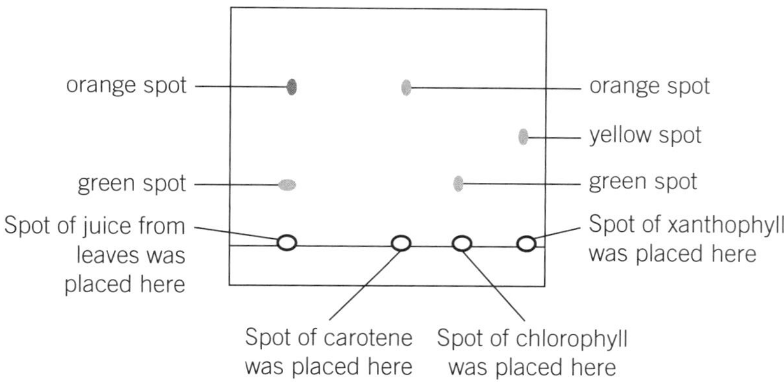

Circle the correct **bold** words in the sentences below.

The chromatogram shows that Sam's leaves contain **one / two / three** pigments. The orange pigment has travelled the same distance as the **carotene / chlorophyll / xanthophyll** reference spot, so it is probably carotene. The

orange / yellow / green pigment has travelled the same distance as the chlorophyll reference spot, so it is

probably chlorophyll. Sam's leaves do not include **carotene / chlorophyll / xanthophyll**.

What you need to remember

The dyes that are mixed together in ink are soluble in the same _______________. This means that you can

separate them by _______________. In this process, the different dyes travel _______________ distances up

the paper. This separates the dyes, which each make a separate _______________ on the chromatogram.

C2 Chapter 2 Pinchpoint 

Pinchpoint question

Answer the question below, then do the follow-up activity **with the same letter** as the answer you picked.

Lilia has two beakers of water. Each contains 100 g of water. She adds 5 g of coffee powder to one beaker and stirs. The coffee powder dissolves to make a solution.

Then Lilia adds 5 g of white sugar to the other beaker and stirs. The sugar dissolves to make a colourless solution.

What are the masses of the two solutions that Lilia makes?

	Mass of coffee solution (g)	Mass of sugar solution (g)
A	100	100
B	105	100
C	95	95
D	105	105

Follow-up activities

A When any substance dissolves in water, its particles are still present in the solution.

Tick the true statements below.

1 When sugar dissolves in water, water particles surround the sugar particles.

2 Sugar particles have no mass when they are dissolved in water.

3 The mass of a sugar particle is the same before and after it dissolves.

4 The total mass of sugar particles decreases when sugar dissolves in water.

5 The mass of sugar solution is equal to the mass of water plus the mass of sugar that dissolves in it.

Hint: If you add sugar to water, there are particles of sugar in the solution. The mass of the sugar particles does not change. The same is true for coffee. For help see C2 2.2 Solutions.

B When any substance dissolves in water, its particles are still present in the solution.

Circle the correct **bold** words and phrases in the sentences below.

Coffee powder dissolves in water to make a **brown / colourless** solution. The coffee particles are **no longer / still** present. The mass of a coffee particle **does not change / decreases** when coffee powder dissolves. This means that the total mass of coffee also **does not change / decreases**. For this reason, the mass of coffee solution is equal to the mass of water **minus / plus** the mass of coffee powder that dissolves in it.

White sugar dissolves in water to make a **brown / colourless** solution. The sugar particles are **no longer / still** present, even though you cannot see the sugar. The mass of a sugar particle **does not change / decreases** and the total mass of sugar also **does not change / decreases**. This explains why the mass of sugar solution is equal to the mass of water **minus / plus** the mass of sugar that dissolves in it.

Hint: If you add coffee powder to water, the total mass of solution is equal to the total mass of water and coffee powder. The same is true for sugar, even though you cannot see the sugar. For help see C2 2.2 Solutions.

C Marcus uses rice and beans to model a sugar solution.

- He has 500 grains of uncooked rice. Each grain of rice represents one water particle.
- He has 10 dried beans. Each bean represents one sugar particle.

Fill in the gaps to complete the sentences about mass and solutions.

Marcus adds 10 beans to 500 grains of rice. This represents adding sugar to water. On mixing, the mass of a grain of

rice does not ___________________ and nor does the mass of a ___________________. This means that the mass of

the mixture is ___________________ to the ___________________ mass of the rice and beans before mixing. Also, the

total number of grains of rice and beans is the ___________________ before and after.

The same rule is true when you make sugar solution. The mass of sugar solution is ___________________ to the total

mass of ___________________ and ___________________ before mixing.

D The mass of a solution is equal to the total mass of solvent and solute.

Calculate the missing masses.

Solution	Mass of solvent (g)	Mass of solute (g)	Mass of solution (g)
W	100	6	
X	200		210
Y		4	100
Z	498		506

Hint: Remember, the mass of solution is equal to the mass of solvent added to the mass of solute. For help see C2 2.3 Solubility.

Pinchpoint review

Now look back at the question – do you think you chose the right letter?
Turn to the Answers page to find out.

C3.1 Acids and metals

A Farai has four test tubes. He half-fills the test tubes with dilute hydrochloric acid.
Then he adds a small piece of a different metal to each test tube.
He observes bubbles of gas in some test tubes.
Circle the names of the **two** metals that make bubbles.

magnesium   **copper**   **zinc**   **gold**

B Some metals react with dilute acids to make hydrogen gas.

Circle the correct **bold** words in the sentences below.

You can use a **lighted / glowing** splint to test for hydrogen gas. Collect some gas in a **beaker / test tube**.

Put a lit splint in the **beaker / test tube**. If the flame goes out with a **squeaky / booming** pop, the gas is

hydrogen. The pop happens because hydrogen reacts with **oxygen / nitrogen** from the air, explosively.

The reaction makes **nitrogen dioxide / water**.

C Magda adds different metals to four test tubes of dilute hydrochloric acid.
The diagram shows her observations.

a Complete the results table for the experiment.

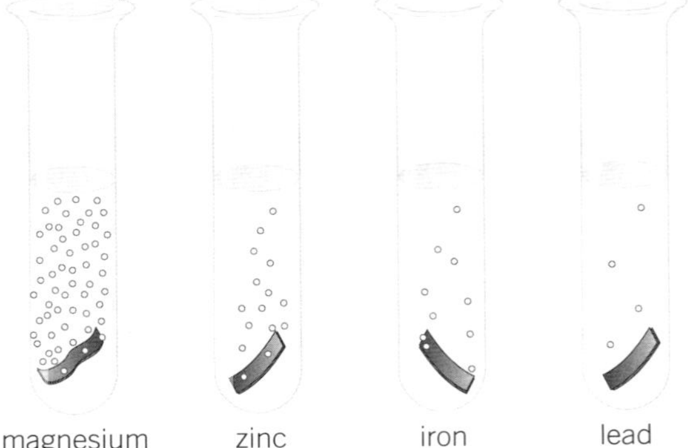

Metal	Observations
magnesium	
zinc	
iron	
lead	A few bubbles formed, very slowly.

b Magda wrote a conclusion for her experiment, but she
made one mistake in each sentence. Use a coloured
pen to correct each mistake.

All the metals reacted with dilute acid, except for one. The reaction with zinc was most vigorous. The reaction
with iron was least vigorous. When we tested the gas in the bubbles, we found out that it was oxygen gas.

D Complete the word equations below.

a magnesium + hydrochloric acid → magnesium chloride + _____________________

b zinc + hydrochloric acid → zinc _____________________ + hydrogen

c iron + hydrochloric acid → _____________________ _____________________ + hydrogen

d lead + hydrochloric acid → _____________________ _____________________ + _____________________

What you need to remember

Most elements are _____________________. They have similar physical properties; for example, they are shiny, and

they conduct heat and _____________________. Metals also have patterns in their chemical properties. For example,

some metals react with dilute acids to make salts and _____________________ gas. Magnesium reacts with dilute

hydrochloric acid to make _____________________ chloride and hydrogen.

C3.2　Metals and oxygen

A　Zion heats small pieces of five metals in a Bunsen burner flame. She writes her observations in the table below. Complete the table with the names of the products.

Metal	Observations when heated in a Bunsen burner flame	Name of product formed when the metal is heated in a Bunsen burner flame
copper	Does not burn. Forms layer of black product on its surface.	copper oxide
iron	Small pieces burn, but less vigorously than zinc.	
lead	Does not burn, but forms a layer of the product on its surface.	
magnesium	Burns vigorously.	
zinc	Small pieces burn, but less vigorously than magnesium.	

B　Here is a conclusion to Zion's experiment in activity **A**. Circle the correct **bold** words.

In the experiment, the metal that burns most vigorously is **zinc / magnesium**. This means that magnesium is the

most / least reactive metal in the experiment. The product of the reaction is **magnesium chloride / magnesium**

oxide. Two metals – lead and copper – **do not / do** burn in air. Instead, they form layers of lead oxide or copper

chloride / oxide on their surface. These are the most **unreactive / reactive** metals in the experiment. Gold is even

less / more reactive than copper and lead.

C　Use the observations in the table in activity **A** to write the names of the metals in order of reactivity, most reactive at the top. Two have been done for you.

Most reactive　　　　magnesium

____

____

____

Least reactive　　　　copper

D　In the reactions in the table in activity **A**, the metals are all in the solid state. They react with oxygen in the gas state. The products are metal oxides. They are in the solid state.
Complete the equations below by writing one state symbol in each set of brackets.

a　$2Mg\ (\quad) + O_2\ (\quad) \rightarrow 2MgO\ (\quad)$

b　$2Cu\ (\quad) + O_2\ (\quad) \rightarrow 2CuO\ (\quad)$

c　$2Zn\ (\quad) + O_2\ (\quad) \rightarrow 2ZnO\ (\quad)$

What you need to remember

Some metals react with oxygen from the ________________. The more ________________ the metal, the more

vigorous the reaction. For example, magnesium burns vigorously to make magnesium ________________.

Copper does not burn when you heat it in a Bunsen burner flame. Instead, it forms a layer of black copper

________________ on its surface. Gold does not react with oxygen. It is an ________________ metal.

State symbols in equations show the ________________ of a substance in the reaction. The state symbol for

________________ is (s) and the state symbol for ________________ is (l).

C3.3 Metals and water

A Mrs Winter adds a small piece of sodium to a large amount of water. The diagram shows the reaction as it happens. Label the diagram using the labels provided.

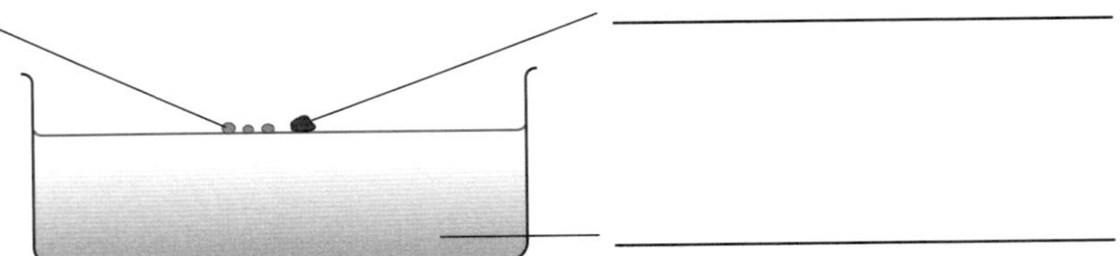

B Look again at the diagram you labelled in activity **A**.
Circle the names of the two reactants in pen.
Circle the names of the two products in pencil.

Hint: One label includes the names of two substances.

C Zara observes the experiment in activity **A**. Mrs Winter then adds lithium and potassium to separate containers of water. Zara writes down her observations.

Metal	Observations
sodium	Sodium whizzes round on the surface of the water. Fizzing.
lithium	Lithium whizzes round on the surface. Fizzing, but less than for sodium.
potassium	Potassium whizzes round on the surface, and there is a bright, lilac coloured flame. Very vigorous fizzing.

Use the observations to write the names of the metals in order of reactivity:

Most 1. _________________ 2. _________________ 3. _________________ **Least**

D The box shows part of the reactivity series.

Draw a line to match each metal with how it reacts with water.

Hint: Use the positions of the metals in the reactivity series to help you.

Metal	Observations on adding to cold water
caesium	Tiny bubbles form very slowly
calcium	No reaction
magnesium	Vigorous reaction, with rapid bubbling
gold	Explosive reaction

caesium
sodium
calcium
magnesium
zinc
iron
copper
gold

What you need to remember

The reactivity series lists the _________________ in order of how vigorously they react. The metals at the _________________ of the reactivity series react vigorously with water, oxygen, and dilute acids. The metals at the bottom of the reactivity series are _________________. Metals that react with water include potassium, sodium, lithium, and calcium. The products of their reactions are _________________ gas and the metal _________________.

C3.4 Metal displacement reactions

A The box shows part of the reactivity series.

Highlight or underline the more reactive metal in each pair below.

Use the reactivity series to help you.

Hint: Remember the metal at the top of the reactivity series is the most reactive.

a Magnesium and gold

b Zinc and aluminium

c Lead and zinc

d Iron and silver

magnesium
aluminium
zinc
iron
lead
copper
silver
gold

B Mohamed sets up the apparatus shown in the diagram.

Circle the correct **bold** words in the sentences below.

The reactivity series shows that zinc is **more / less** reactive than copper. This

means that zinc can **oxidise / displace** copper from copper sulfate solution.

The products of the reaction are **copper / zinc**, which forms on the surface of

the zinc, and zinc **sulfate / chloride** solution.

C Tick the pairs of substances in the table that are likely to react in displacement reactions, assuming the conditions are suitable.

1 Magnesium and lead oxide ☐

2 Zinc and copper sulfate solution ☐

3 Copper and zinc chloride solution ☐

4 Lead and iron oxide ☐

5 Zinc and lead nitrate solution ☐

D Complete the word equations for the displacement reactions below.

a magnesium + copper sulfate solution → _____________ _____________ solution + copper

b aluminium + lead oxide → aluminium oxide + _____________

c zinc + copper chloride solution → zinc chloride solution + _____________

d magnesium + iron oxide → _____________ oxide + _____________

What you need to remember

Metals take part in displacement reactions. In a displacement reaction, a _____________ reactive metal

pushes out a _____________ reactive metal from its compound. For example, in the thermite reaction,

aluminium displaces iron from _____________ _____________. The products of the reaction are

aluminium _____________ and _____________.

C3.5 Extracting metals

A Most metals are found in the Earth as compounds. A few are found as elements.

Circle the correct **bold** words in the sentences below.

An element is a substance that **can / cannot** be broken down into other substances. All the atoms in an element

are **the same / different**. A compound is a substance made up of atoms of two or **more / fewer** elements. The

atoms of the different elements in a compound are **not / strongly** joined together.

B The box shows part of the reactivity series.

> sodium
> magnesium
> aluminium
> carbon
> zinc
> iron
> lead
> copper

   a Circle the name of the **non-metal** element in the reactivity series.

   b In the list below, circle the names of the metals that can be extracted from their compounds by heating with the non-metal from part **a**.

      aluminium   **copper**   **iron**   **lead**   **magnesium**   **zinc**

   c Suggest why you cannot extract aluminium from its compounds by heating the compounds with carbon.

     ____

C An ore is a rock that it is worth extracting a metal from.
The statements below can be reordered to describe how to get iron from iron ore.
Read the statements and write down the order of statements you think will give the best description.

Correct order   ☐ ☐ ☐

   1 Separate iron oxide from compounds that are mixed with it.

   2 Dig the iron ore out of the ground.

   3 Use a displacement reaction to extract iron from iron oxide.

D An ore from one zinc mine is 20% zinc.

   a Calculate the percentage of the ore that is **not** zinc.

   b Calculate the mass of zinc in 100 kg of ore.

   c Calculate the mass of zinc in 2000 kg of ore.

What you need to remember

Most metals exist in the Earth's _________________ as compounds. These compounds are _________________

with other compounds in rock. A rock that it is worth extracting a metal from is called an _________________.

Many metals are extracted from their compounds in _________________ reactions. For example, iron oxide

is heated with carbon. The carbon _________________ iron from iron oxide. The products of the reaction are

_________________ and carbon dioxide.

C3.6 Ceramics

A Tick the statements below that are true.

1   Pottery is a ceramic material. ☐

2   Ceramics are brittle, so they break if you drop them. ☐

3   Ceramics are good conductors of electricity. ☐

4   Ceramics are hard. ☐

5   Ceramics react vigorously with acids. ☐

B The table lists some uses of ceramics. Tick **one or more** boxes next to each use to show the property or properties that make a ceramic suitable for this use.

Use	Property or properties that make ceramics suitable for this use			
	Do not conduct electricity	Do not react with water	High melting point	Strong when forces press on it
Making walls				
Making jet engine blades, which get very hot				
Making plates and cups				
Insulators for overhead power lines				

C Ruby sets up the apparatus below to compare the strengths of different ceramic materials. She finds the mass that makes the piece of ceramic break.

For each variable listed below, identify which type of variable it is in this experiment.

independent variable	control variable	dependent variable

The mass that makes the ceramic break. ___________________________________

The position of the masses. _____________________________

The ceramic material. ___________________________

The size / thickness of the piece of ceramic material. _______________________________

What you need to remember

Ceramics are compounds. They include metal silicates, _______________ oxides, and _______________ carbides. Ceramics are hard, stiff, and brittle. They have very _______________ melting points and are electrical _______________. These are _______________ properties. Ceramics also have similar chemical properties to each other – they do not react with water, acids, or _______________.

C3.7 Polymers

A The sentences below are about polymers. There is one mistake in each sentence.
Copy out each sentence, correcting its mistake.

a A polymer is a substance with very short molecules.

____

b A polymer molecule has many different groups of atoms, joined together in a long chain.

____

c There are 10 different polymers.

____

d All polymers have very similar properties.

____

B Draw a line from each use of a polymer to a property that makes it suitable for this use.

| Poly(propene) is used to make ropes |

| because it is strong and flexible, does not wear away, and is not damaged by blood. |

| Nylon can be used to make artificial heart valves |

| because it is strong and lightweight. |

| Cotton is used to make clothes |

| because it can be spun into flexible threads that can be woven into cloth. |

| Kevlar® is used to make bullet-proof vests that police officers can wear all day |

| because it is strong and flexible. |

C The table gives data for three polymers.
Choose the **two** polymers in the table that are suitable for making water bottles.
Give a reason for your choice.

Name of polymer	Is it waterproof?	Density (g/cm³)	Flexibility
LDPE	yes	0.92	very flexible
PETE	yes	1.38	rigid
Rigid PVC	yes	1.30	rigid
Poly(propene)	yes	0.90	very flexible

Polymer names _________________________________ and _________________________________

Reason ____

____

What you need to remember

A polymer is a substance with very _________________ molecules. Its molecules have identical groups of atoms,

repeated _________________ times. There are _________________ polymers, each with _________________

properties. The properties of polymers depend on the groups of _________________ in their molecules.

C3.8 Composites

A Tick the statements below that are true.

   1   A composite material is a mixture of two or more materials. ☐

   2   The properties of a composite material are the same as the properties of one of the materials that is in it. ☐

   3   Reinforced concrete is made of steel bars that are surrounded by concrete. ☐

   4   Reinforced concrete is used to make tall buildings. ☐

   5   Reinforced concrete is stronger than concrete if you hang weights on it. ☐

B **a**   Draw a line from each object to the properties its materials must have.

| aeroplane wing |
| heat shields for spacecraft |
| hockey stick |
| brake discs in high-performance cars |

| light and strong |
| not damaged at high temperatures |

   b   Complete the sentences below. Use your answers to part **a** to help you.

   i   Glass-fibre-reinforced aluminium is used to make aeroplane wings because it is

   ____

   ii   Carbon-fibre-reinforced carbon is very expensive. It keeps its strength up to temperatures of 2000 °C, so it can be used to make

   ____

   iii   Carbon-fibre-reinforced silicon carbide is expensive. It is not damaged at high temperatures. It is heavier for its size than carbon-fibre-reinforced carbon. This makes it suitable for

   ____

C A team of scientists measured the strength when poly(propene) was reinforced by different fibres. The table shows some of their data.

Fibre	Strength when pulled (MPa)
silk	47
wood	32
bamboo	35
glass	86

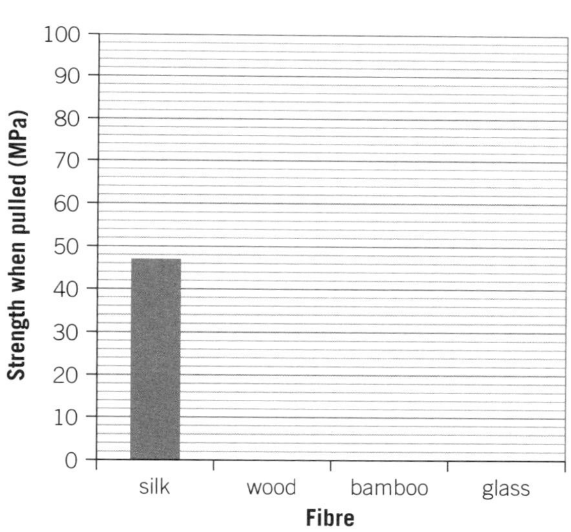

   a   Plot the data from the table on a bar chart, using the axes provided.

   b   Write two sentences to describe what your bar chart shows.

   ____

   ____

Pinchpoint question

Answer the question below, then do the follow-up activity **with the same letter** as the answer you picked.

Which pair of substances take part in a displacement reaction, and why?

Use the reactivity series shown to help you decide.

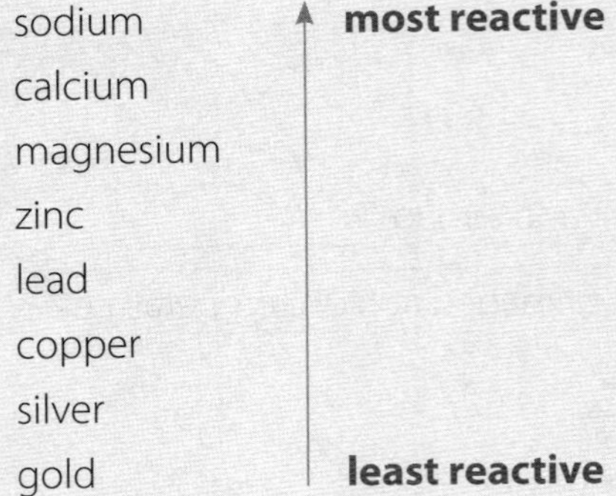

A   Lead and zinc nitrate solution because lead is more reactive than zinc.

B   Copper and lead nitrate solution because lead is more reactive than copper.

C   Zinc and copper nitrate solution because zinc is more reactive than copper.

D   Magnesium and sodium nitrate solution because magnesium and sodium are both very reactive.

Follow-up activities

A   In a displacement reaction, a more reactive metal displaces, or pushes out, a less reactive metal from its compound. A metal is more reactive than another if it is higher in the reactivity series.

Circle the **more** reactive metal in each pair. Use the reactivity series above to help you.

a   lead and copper

b   magnesium and sodium

c   zinc and lead

d   sodium and zinc

e   zinc and magnesium

f   calcium and magnesium

g   gold and silver

Hint: A metal is more reactive than another if it is higher in the reactivity series. Which is more reactive, lead or copper? For help see C2 3.4 Metal displacement reactions.

B A displacement reaction occurs when a more reactive metal displaces, or pushes out, a less reactive metal from its compound.

  a Tick the **four** true statements below. Use the reactivity series to help you.

  1 Copper is more reactive than silver. ☐

  2 Copper displaces silver from its compounds. ☐

  3 Lead is more reactive than zinc. ☐

  4 Zinc displaces lead from its compounds. ☐

  5 Magnesium is more reactive than zinc. ☐

  6 Zinc displaces magnesium from its compounds. ☐

  7 Magnesium displaces calcium from its compounds. ☐

  8 Gold displaces silver from its compounds. ☐

  b Write out corrected versions of the **four** incorrect statements.

  ____

  ____

  ____

  ____

Hint: The higher a metal is in the reactivity series, the more reactive it is. Which metal is more reactive, copper or silver? For help see C2 3.4 Metal displacement reactions.

C Circle the correct **bold** words and phrases in the sentences below.

Magnesium and zinc oxide **react / do not react** in a displacement reaction. This is because magnesium is **more / less** reactive than zinc.

Lead **displaces / does not displace** copper from copper sulfate solution. This is because lead is **more / less** reactive than copper.

Zinc **displaces / does not displace** magnesium from magnesium nitrate solution. This is because zinc is **more / less** reactive than magnesium.

Zinc **displaces / does not displace** sodium from sodium chloride solution. This is because sodium is **more / less** reactive than zinc.

Sodium **displaces / does not displace** calcium from calcium nitrate solution. This is because sodium is **more / less** reactive than calcium.

Gold **displaces / does not displace** lead from lead nitrate solution. This is because gold is **more / less** reactive than lead.

Hint: A metal displaces another metal from its compound if the metal on its own is higher in the reactivity series than the metal in the compound. Which metal is more reactive, magnesium or zinc? For help see C2 3.4 Metal displacement reactions.

D A metal is more reactive than another if it is higher in the reactivity series. In a displacement reaction, a more reactive metal displaces, or pushes out, a less reactive metal from its compound.

a Circle the name of the more reactive metal in each pair of metals. Use the reactivity series above to help you.

magnesium and copper

copper and lead

zinc and lead

lead and magnesium

magnesium and sodium

zinc and magnesium

Hint: Look at the reactivity series to find out which metal is more reactive. For help see C2 3.4 Metal displacement reactions.

b Look at the pairs of substances in each shape below. Colour in green the shapes in which the substances **do react**. Colour in red the shapes in which the substances **do not react**.

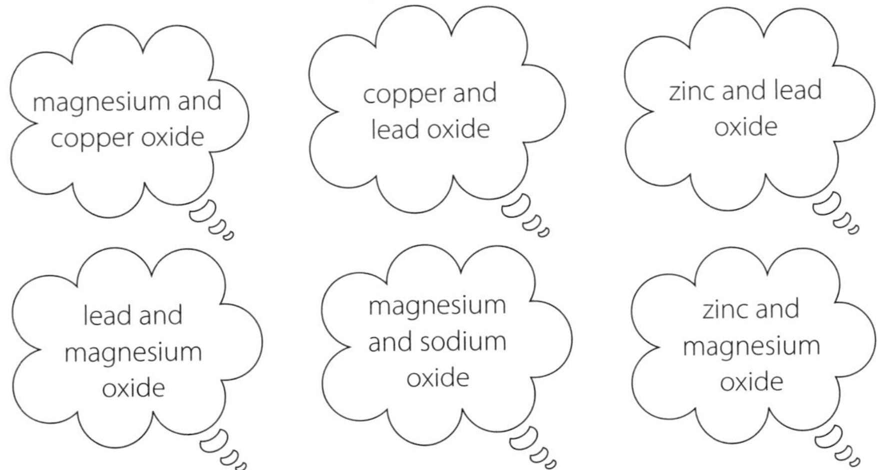

Hint: In a displacement reaction, a more reactive metal displaces, or pushes out, a less reactive metal from its compound. For help see C2 3.4 Metal displacement reactions.

Pinchpoint review

Now look back at the question – do you think you chose the right letter?
Turn to the Answers page to find out.

C4.1 The Earth and its atmosphere

A The diagram shows a cross-section of the Earth. Label the diagram using the key words provided.

> **crust**   **inner core**   **mantle**   **outer core**

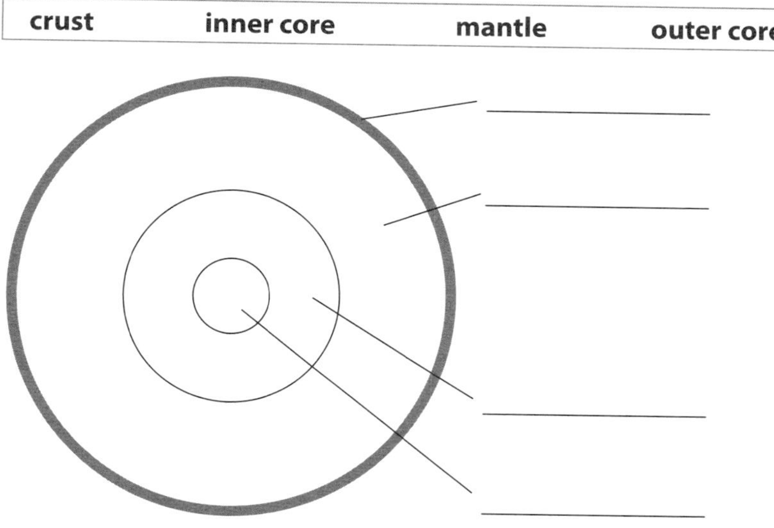

B Draw a line from each layer of the Earth to its description.

crust	mainly solid rock, but can flow
mantle	liquid, mainly iron and nickel
inner core	rocky, of thickness 8 km to 40 km
outer core	solid, mainly iron and nickel

C You can use a hard-boiled egg in its shell to model the structure of the Earth. The statements below describe the egg.

Write **S** next to each statement that describes a **similarity** of the egg to the Earth.

Write **D** next to each statement that describes a **difference** of the egg to the Earth.

1 The layer on the outside is thin ☐ 

2 It is made up of separate layers ☐ 

3 The layer in the centre is all solid ☐ 

4 It is egg-shaped ☐ 

5 The layer on the outside is solid ☐ 

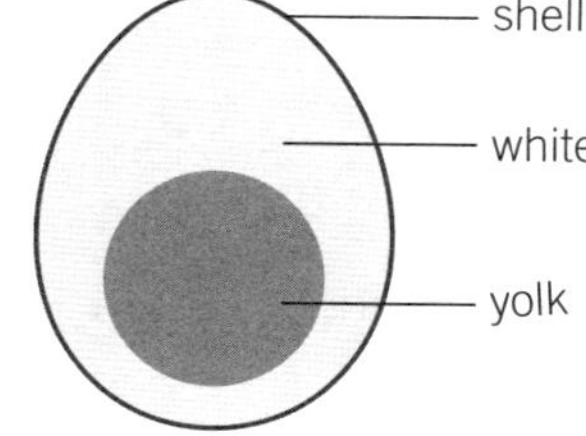

What you need to remember

The Earth is made up of four _______________. In the centre is the solid _______________ core, which is surrounded by the liquid _______________ core. Outside the outer core is the _______________. This is mainly _______________ rock, but it can _______________. On the surface of the Earth is the rocky _______________, which is between 8 km and 40 km thick. The gases of the _______________ surround the Earth. The layer of the atmosphere nearest the Earth is the _______________. This is a mixture of gases, including 78% _______________ and 21% _______________.

C4.2 Sedimentary rocks

A Circle the correct **bold** words in the sentences below.

Sedimentary rocks are made up of separate grains that are joined together. There are **huge / tiny** gaps between the grains. Air and water can get into these gaps. This explains why sedimentary rocks are **porous / soft**.

The forces holding the grains together are **not very strong / very strong**. This explains why sedimentary rocks are **porous / soft**.

B The statements below can be reordered to describe how sedimentary rocks are made. Read the statements and write down the order of statements you think will give the best description. Start with statement number 1.

Correct order ☐ ☐ ☐ ☐

1 Weathering breaks up all types of rock into much smaller pieces or sediments.

2 The sediments stop moving and settle in one place in a process called deposition.

3 The sediments join together to make new rock.

4 These small sediments are transported far away from the rock.

C Draw a line from each key word to its definition.

physical weathering	The breaking of a rock into sediments, and their movement away.
erosion	The breaking up of rock by the action of substances, for example those in rainwater.
chemical weathering	The 'gluing together' of sediments by different substances to make sedimentary rocks.
deposition	The settling of sediments that have moved away from their original rock.
cementation	The breaking up or wearing down of rock because of changing temperature, for example.
compaction	Squashing sediments together to make new rock by the weight of layers above.
transport	The movement of sediments far away from their original rock.

What you need to remember

There are three types of rock – igneous, sedimentary, and ______________. Most types of sedimentary rock are porous and ______________. Sedimentary rocks are formed when rock breaks up into smaller pieces by physical, chemical, or ______________ weathering. The smaller pieces, called ______________, move away from the rock and are carried far away by ______________ processes. The sediments settle in one place. This is ______________. Then the sediments join together by ______________ or ______________.

In ______________, the weight of the sediments above ______________ together the sediments below. In ______________, another substance sticks the sediments together.

A For each rock name below, give whether it is an igneous, a metamorphic, or a sedimentary rock.

basalt _________________________   marble _________________________

granite _________________________   sandstone _________________________

limestone _________________________   slate _________________________

B Tick **one or two** boxes next to each statement to show whether the statement is true for igneous rocks or metamorphic rocks, or both.

	Statement	✓ if true for igneous rocks	✓ if true for metamorphic rocks
a	They are made up of crystals.		
b	They are not porous.		
c	They form when liquid rock (magma or lava) cools and solidifies.		
d	They form when heat and/or high pressure change existing rocks.		
e	Their formation involves a change of state.		

C Salol is a substance that has a melting point of 42 °C. When hot liquid salol cools down in the lab, it freezes to make crystals.

Maddie has some liquid salol. She places a few drops on a cold microscope slide, and a few drops on a warm microscope slide. She obtains the results shown below.

Circle the correct **bold** words in the sentences below.

On both microscope slides, the liquid salol **freezes / evaporates**

to make solid salol. The salol on the warm slide cools more

slowly / quickly. Its particles have **less / more** time to arrange

themselves into crystals, so the crystals are **smaller / bigger**. When liquid rock cools and freezes slowly, the crystals

in the **metamorphic / igneous** rock that is formed are **small / big**. When liquid rock cools and freezes quickly, the

crystals in the **metamorphic / igneous** rock that is formed are **small / big**.

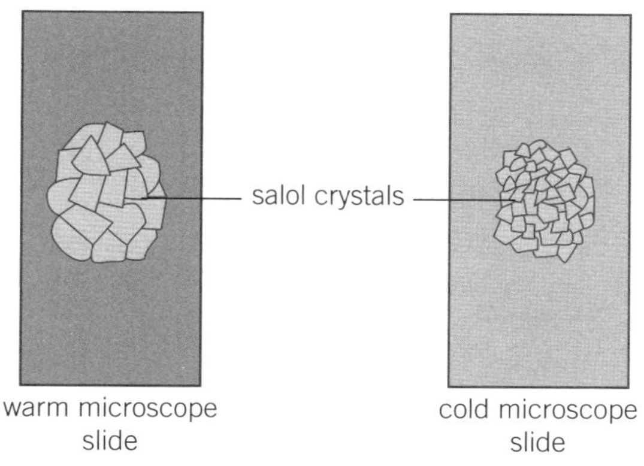

What you need to remember

Underground, liquid rock is called _________________. On the surface, liquid rock is called _________________.

Igneous rocks form when liquid rock cools and _________________. They are made up of _________________,

which are joined together with no gaps. This means that igneous rocks are ___________-___________. They are also

durable and _________________. Metamorphic rocks form when heat and/or high _________________ change

existing rock. Metamorphic rocks are made up of _________________, so they are non-porous.

C4.4 The rock cycle

A Draw lines to make five correct sentences about how the materials in rocks are recycled. Use each sentence part once, more than once, or not at all.

Solid rock of any type		freezes		to make sediments.
		melts		to make lava or magma.
Liquid rock		is broken into smaller pieces		to make igneous rock.
		join together		to make metamorphic rock.
Sediments of any rock type		has its particles rearranged		to make sedimentary rock.

B Each number below represents a process that helps to convert rock from one type to another.

1	cooling	**6**	weathering
2	freezing	**7**	transport
3	melting	**8**	deposition
4	the action of heat	**9**	cementation
5	the action of high pressure	**10**	compaction

Label the diagram of the rock cycle by writing **one or more** numbers in each box. Use each number as many times as you need to.

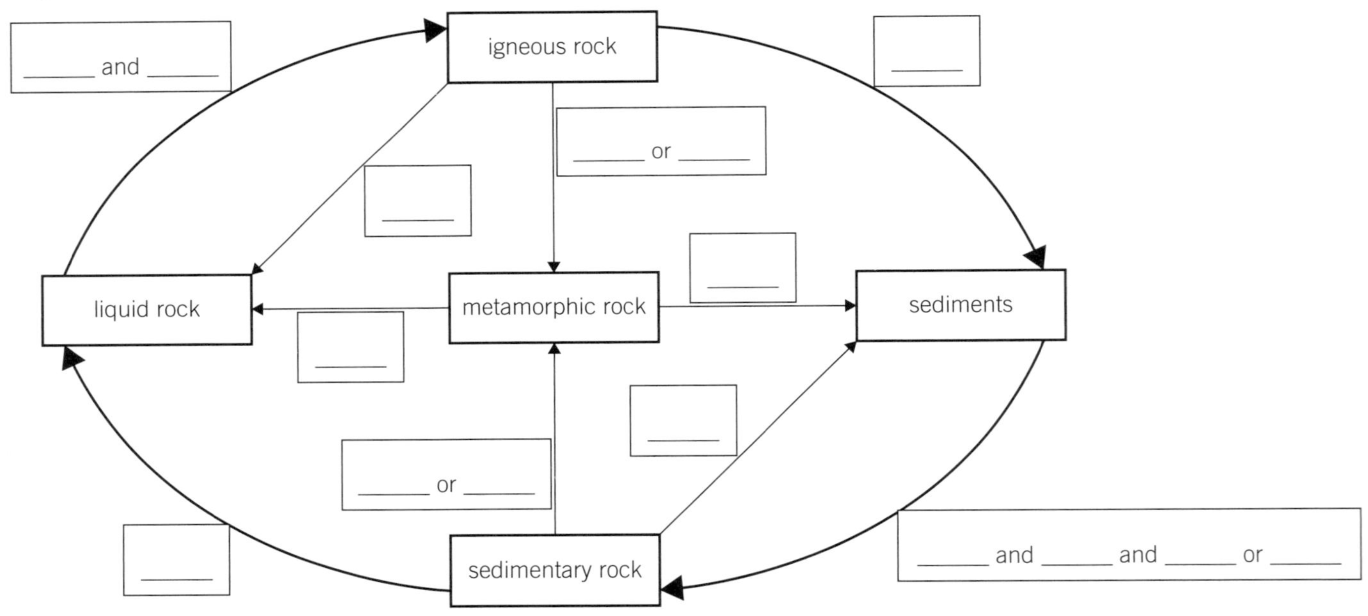

What you need to remember

The rock cycle shows how the materials in rocks are _____________________. For example, when a rock of any type

is weathered, its sediments may form __________________ rocks. When a rock melts, the liquid rock later cools

and _________________ to make an __________________ rock. When a rock experiences high pressures or

__________________, its particles may be rearranged, forming a __________________ rock. When forces from inside

the Earth push rocks upwards, __________________ occurs.

C4.5 The carbon cycle

A Highlight or underline the stores (or reservoirs) of carbon in the list below.

oceans   **fossil fuels**   **all igneous rocks**

the atmosphere   **solar panels**   **some sedimentary rocks**

B Circle the correct **bold** words in the sentences below.

Carbon dioxide is entering and leaving the atmosphere all the time. For many years, carbon dioxide was added to the atmosphere at the same rate as it left the **oceans / atmosphere**. The concentration of carbon dioxide **did not change / increased**. More recently, the concentration of carbon dioxide in the atmosphere has **decreased / increased**. This is because it is added to the atmosphere **faster / slower** than it is removed.

C Colour the processes that **add** carbon dioxide to the atmosphere red.
Colour the processes that **remove** carbon dioxide from the atmosphere green.

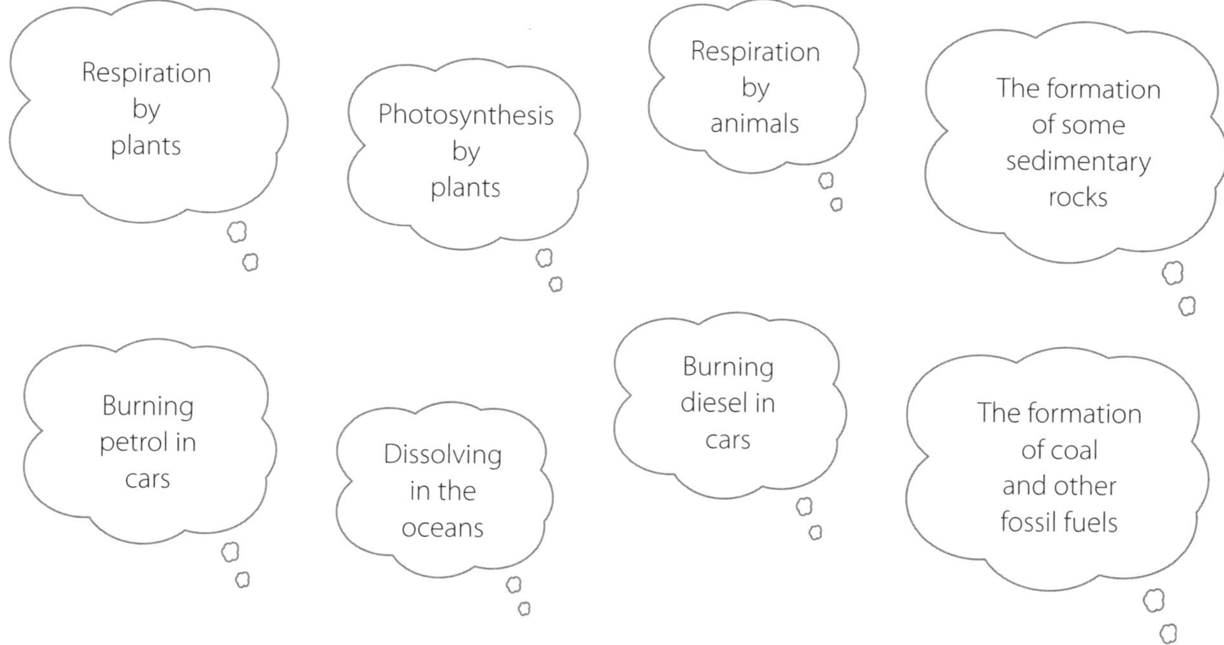

What you need to remember

There are several carbon stores, or ________________. These include the atmosphere, the ocean,

________________ rocks, ________________ fuels, plants and animals, and the soil. The carbon

________________ shows how carbon atoms move between stores. For example, carbon dioxide enters the

atmosphere when plants and animals ________________. It also enters the atmosphere when fossil fuels

________________. Carbon dioxide leaves the atmosphere when plants use it in ________________. It also

leaves the atmosphere by ________________ in oceans. Before industrialisation, carbon dioxide was added to the

atmosphere at the ________________ rate as it left the atmosphere. This meant that the concentration of carbon

dioxide did not ________________.

C4.6 Climate change

A Complete the table to match each word or phrase with its definition. Choose from the list below.

climate change	global warming	greenhouse effect	deforestation

Word or phrase	Definition
	an increase in global average temperatures
	changes to long-term weather patterns
	cutting down or burning trees
	gases in the atmosphere keeping the Earth warmer than it would be if the gases were not there

B Draw a line to match each cause to one direct effect.

Cause	Effect
Deforestation	More carbon dioxide goes into the atmosphere
Burning fossil fuels	The concentration of carbon dioxide in the atmosphere increases
Every year, more carbon dioxide is added to the atmosphere than is removed	Less carbon dioxide is removed from the atmosphere
Climate change	Rising sea levels, floods, and more hurricanes and droughts

C Reorder the following statements to explain how climate change occurs.

Correct order ☐ ☐ ☐ ☐ ☐ ☐

1 The Sun emits radiation, which reaches the Earth.

2 This extra carbon dioxide in the atmosphere means that more of the radiation that is reflected from the surface is absorbed by the atmosphere. This warms the atmosphere by a greater amount than before.

3 Some of this radiation is absorbed by gases in the atmosphere, such as carbon dioxide.

4 Human activities add extra carbon dioxide to the atmosphere.

5 This causes global warming, which is an increase in global average temperatures. Global warming results in climate change.

6 Some of the radiation from the Sun is absorbed by the Earth. The rest of the radiation is reflected by the surface of the Earth.

What you need to remember

Since 1800, _____________ carbon dioxide has been added to the atmosphere than has been removed. This is partly because humans burn _____________ fuels. It is also because humans cut down lots of trees. This is _____________. Carbon dioxide in the atmosphere absorbs some of the solar _____________ reflected off the Earth's surface. This keeps the Earth _____________ than it would otherwise be. This is called the _____________. Extra carbon dioxide results in _____________ average global temperatures, which is _____________. This causes _____________ change, leading to rising sea levels and more extreme _____________ events.

C4.7 Recycling

A Highlight or underline the **two** examples of recycling in the list below.

1 Collecting and melting plastic shampoo bottles to make football shirts.

2 Washing out a used glass jar and keeping coins in it.

3 Collecting and melting gold from old mobile phones to make jewellery.

B The statements below can be reordered to describe how aluminium is recycled. Read the statements and write down the order of statements you think will give the best description.

Correct order ☐ ☐ ☐ ☐ ☐

1 Use a lorry to collect used cans.

2 Pour the liquid aluminium into a mould.

3 Leave the liquid aluminium to cool and freeze.

4 Melt the shreds of aluminium in a furnace.

5 Shred the cans.

C Aluminium may be obtained from its ore, or by recycling used aluminium objects. Tick one box next to each statement to show whether the statement is describing an advantage or a disadvantage of recycling.

	Statement	✓ if this is an advantage of recycling	✓ if this is a disadvantage of recycling
1	Recycling uses less energy.		
2	Some people do not like sorting their waste.		
3	The world's aluminium ore will last longer.		
4	Recycling makes less waste.		
5	The lorries that collect cans for recycling make polluting gases.		

D Different plastics must be sorted before recycling. If a plastic is less dense than water, it floats. If a plastic is more dense than water, it sinks. The bar chart shows the densities of four plastics. The density of water is 1.0 g/cm³.

Write down the names of the **two** plastics in the bar chart that float on water.

_______________________ , _______________________

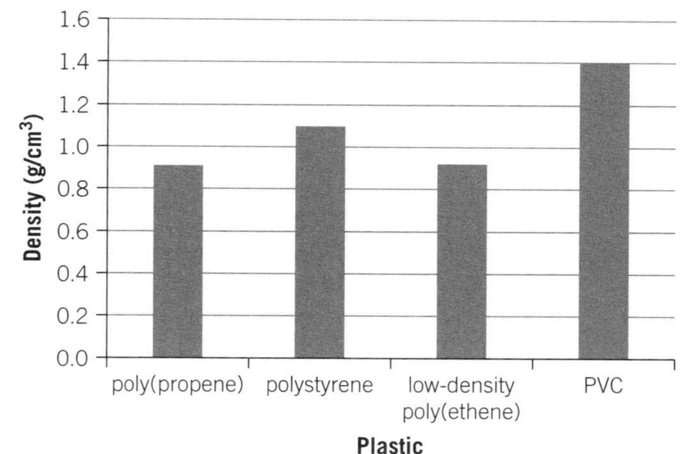

What you need to remember

Recycling means collecting and _________________ used materials so that they can be used again. There are

many advantages of recycling, including _________________ energy needs. Disadvantages of recycling include the

_________________ produced by lorries that collect objects for recycling.

Pinchpoint question

Answer the question below, then do the follow-up activity **with the same letter** as the answer you picked.

Which statement is the **best** description of one way in which metamorphic rock is formed from sedimentary rock?

A Sedimentary rock experiences high pressure, and its particles are rearranged.

B The particles in sedimentary rock are rearranged.

C Existing rock is broken up by weathering. This makes small pieces of rock, which join together by cementation.

D Sedimentary rock experiences high temperatures. It melts, and its particles are rearranged.

Follow-up activities

A Draw a line from each description of processes to show the type of rock that is formed as a result of these processes.

Description of processes

Rock is broken up to make small pieces of rock. The small pieces of rock move away and are deposited. They join together by compaction or cementation.

Rock is heated. It does not melt, but its particles are rearranged.

Rock is heated. It melts and freezes again.

Rock experiences high pressures. Its particles are rearranged.

Type of rock that is formed

igneous

metamorphic

sedimentary

Hint: The formation of both igneous and metamorphic rock involves heating, but which is formed from liquid rock? For help see C2 4.3 Igneous and metamorphic rocks and C2 4.4 The rock cycle.

B The sentences below can be reordered to describe how metamorphic rock is formed.
Read the statements and write down the correct order of statements you think will give the best explanation.

Correct order ☐ ☐ ☐ ☐ ☐

1 The action of heat or pressure changes the arrangement of particles in the rock, so forming crystals.

2 An area of hot rock underground heats up some existing rock…

3 This new rock is called metamorphic rock.

4 …and/or the action of high pressure squashes some existing rock.

5 This rock has a new, different structure, so its properties are different from the sedimentary rock it was formed from.

Hint: Metamorphic rock can form as a result of the action of heat or high pressure on existing rock. For help see C2 4.3 Igneous and metamorphic rock.

C Write **M** next to the statements that are true of metamorphic rock, **S** next to the statements that are true of sedimentary rock, and **B** next to the **one** statement that is true of both metamorphic and sedimentary rock.

1 The first step in making this type of rock is to break up existing rock by weathering.

2 It is made when existing rock experiences high pressures.

3 It is made up of small pieces of broken rock.

4 It is made when existing rock experiences high temperatures.

5 Its formation involves small pieces of rock joining together by compaction or cementation.

6 Its formation involves particles being rearranged to form regularly arranged crystals.

7 Its formation does not involve melting.

Hint: The formation of sedimentary rock involves the breaking up of rock into sediments, which then join together. What is involved in the formation of metamorphic rock? For help see C2 4.2 Sedimentary rock and C2 4.3 Igneous and metamorphic rock.

D Circle the correct **bold** words and phrases in the sentences below.

Metamorphic rock can be formed in two ways. One way involves an existing rock getting **hot / cold**. The rock

melts / does not melt, but its particles rearrange themselves to make the **sediments / crystals** of a metamorphic

rock.

The other way of making metamorphic rock involves an existing rock experiencing **low / high** pressures. This also

makes the particles rearrange themselves to form the **crystals / sediments** of a metamorphic rock.

Hint: Does the formation of a metamorphic rock involve melting, or not? For help see C2 4.3 Igneous and metamorphic rocks.

Pinchpoint review

Now look back at the question – do you think you chose the right letter?
Turn to the Answers page to find out.

C2 Revision questions

1 Which gas is present in the largest amount in the Earth's atmosphere? Tick **one** box. *(1 mark)*

argon ☐   carbon dioxide ☐

nitrogen ☐   oxygen ☐

2 Mrs Allfrey adds a small piece of sodium to a large amount of water. She observes vigorous bubbling.

What are the products of the reaction?
Tick **two** boxes. *(2 marks)*

hydrogen ☐   oxygen ☐

sodium chloride ☐   sodium hydroxide ☐

3 Draw **one** line from each mixture to the technique you could use to separate the named substance from the mixture. *(3 marks)*

Substance and mixture	Technique
salt from salt solution	chromatography
water from salt solution	distillation
	evaporation
undissolved salt from saturated salt solution	filtration

4 Neon is an element. It is in Group 0 of the Periodic Table. Which property does neon have?
Tick **one** box. *(2 marks)*

Vigorous reactions ☐   High melting point ☐

High density ☐   Low boiling point ☐

5 **Table 1** shows properties for three elements in the same group of the Periodic Table. The elements are in the same order in the Periodic Table.

Table 1

Element	Density (g/cm³)	Melting point (°C)
Manganese	7	1240
Technetium	11	2200
Rhenium	21	3180

a Plot the **density** data on a bar chart. Use the axes below. *(1 mark)*

b Describe the pattern in the **melting point** data in **Table 1**. *(1 mark)*

____

____

6 Draw **one** or **two** lines from each rock type to show how it may be formed. *(3 marks)*

Rock type	How it may be formed
	by high pressures changing existing rock
igneous	by joining together small pieces of solid rock
metamorphic	by freezing liquid rock
	by high temperatures changing existing rock, without melting

7 **Table 2** shows some properties of three polymers.

Table 2

Polymer	Properties
Low-density poly(ethene)	• flexible • absorbs some liquids
Nylon 6.6	• absorbs water • reacts with sulfuric acid
Poly(propene)	• stiff and rigid • does not absorb water • does not react with acids or solvents

Write down the name of **one** polymer from the table that can be used to make fuel tanks for cars. *(1 mark)*

____

8   **Figure 1** shows a chromatogram of a mixture, and of four pure substances, **W**, **X**, **Y**, and **Z**.

Figure 1

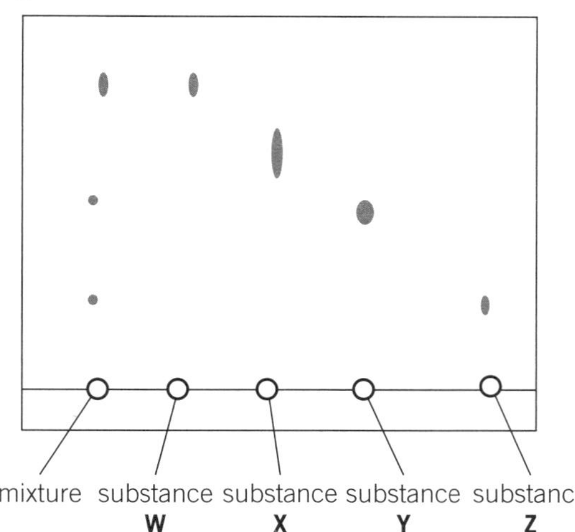

Write down the letters of the **three** substances that are in the mixture. (1 mark)

____

9   Poppy dissolves sugar in water to make a mixture. Draw **one** line from each substance to the word that describes it in the mixture. (1 mark)

Substance	**Word that describes the substance in the mixture**
	saturated
sugar	solute
water	solution
	solvent

10   **Table 3** shows the properties of four elements. The letters of the elements in the table are not the chemical symbols of the elements.

Table 3

Element	Melting point (°C)	Does it conduct electricity?	Appearance at room temperature
W	3730	no	colourless, sparkly
X	1675	yes	grey, shiny
Y	113	no	yellow, not shiny
Z	850	yes	silver colour, shiny

Write down the letters of **two** elements in the table that are metals. (2 marks)

_____________ and _______________

11   Angelina has 85 g of water. She adds salt to the water to make a solution. The mass of the solution is 96 g. Calculate the mass of salt Angelina added to the water. (1 mark)

____

12   Mohamed has a mixture of sand and water. Describe how he can obtain water from the mixture. In your answer, include the names of the pieces of apparatus he needs. (6 marks)

____

____

____

____

____

____

____

____

____

13   Look at **Figure 2**. Which of the diagrams **A** to **D** shows a mixture of substances? (1 mark)

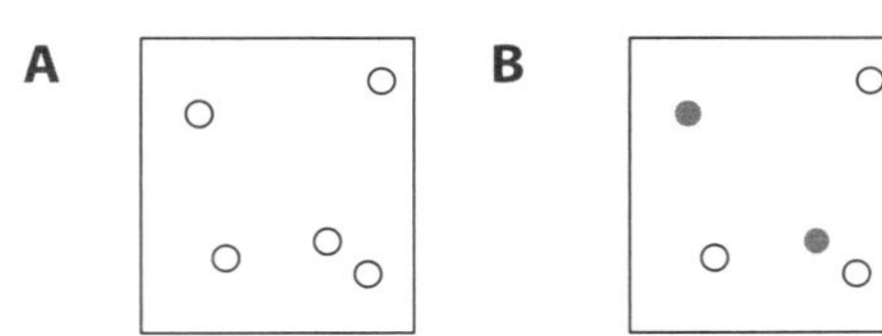

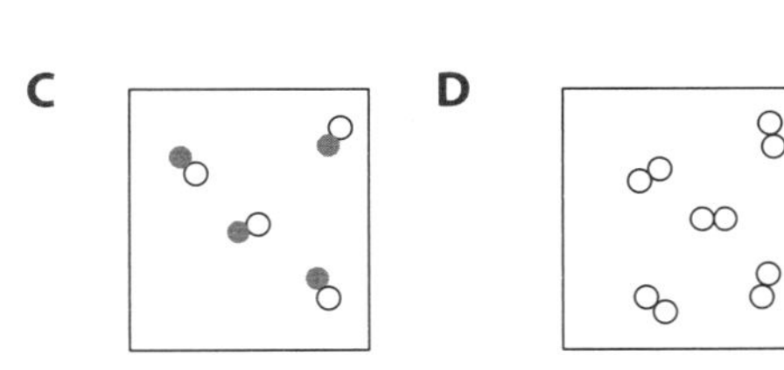

Figure 2

____

14   Scarlett has small pieces of three different metals – **X**, **Y**, and **Z**. She wants to see which metal is most reactive.

She adds each piece of metal to some dilute hydrochloric acid.

Table 4 shows her results.

Table 4

Metal	How quickly bubbles are formed
X	very quickly
Y	no bubbles formed
Z	quite slowly

a Write down the letter of the **one** metal that could be magnesium. *(1 mark)*

b What is the **independent** variable in the reaction? Tick **one** box. *(1 mark)*

Type of metal ☐

How quickly bubbles are formed ☐

Temperature ☐

Volume of acid ☐

c Write down **two** control variables in Scarlett's experiment. *(2 marks)*

15 Some iron ore from a mine contains 35% iron by mass. Calculate the mass of iron in 200 kg of the ore. *(2 marks)*

16 Which word equations show displacement reactions?
Tick **two** boxes. *(2 marks)*

1 iron + bromine → iron bromide ☐

2 sodium + chlorine → sodium chloride ☐

3 chlorine + sodium bromide →
sodium chloride + bromine ☐

4 bromine + potassium iodide →
potassium bromide + iodine ☐

17 The graph in **Figure 3** shows the solubility of sodium nitrate at different temperatures.

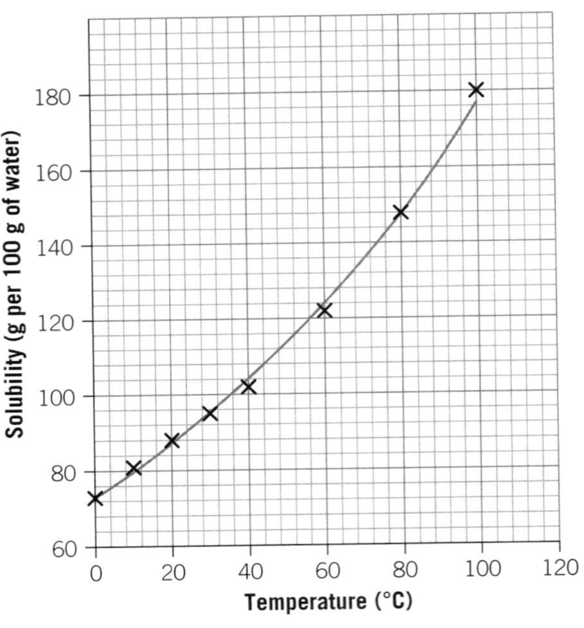

Figure 3

a Write down the meaning of **solubility**. *(1 mark)*

b Describe the pattern shown on the graph. *(1 mark)*

c Use the graph to find the solubility of sodium nitrate at **i** 30 °C and **ii** 50 °C. *(2 marks)*

i _______________ **ii** _______________

18 Explain why the concentration of carbon dioxide in the atmosphere did not change for many years. Use these words in your answer: *(3 marks)*

dissolving	**photosynthesis**	**respiration**

C2 Checklist

Revision question number	Outcome	Topic reference	😦	😐	😊
1	Name the main components of the atmosphere.	C2 4.1			
2	State the products of the reaction between a Group 1 metal and water.	C2 1.3			
3	State some mixtures that can be separated using evaporation.	C2 2.5			
	State some mixtures that can be separated using distillation.	C2 2.5			
	State some situations in which filtering is used.	C2 2.4			
4	State a physical property of Group 0 elements.	C2 1.5			
5a	Plot data on a bar chart.	C2 1.2 WS 1.4			
5b	Describe trends shown by bar chart data.	C2 1.2 WS 1.4			
6	State simply how igneous and metamorphic rocks are formed.	C2 4.3			
7	Identify a suitable polymer to use when given simple information about the polymer.	C2 3.7			
8	Analyse chromatograms to identify substances in mixtures.	C2 2.6			
9	Describe solutions using key words.	C2 2.2			
10	Use observations about elements to decide if they are metals or non-metals.	C2 1.1			
11	Use data to predict how much solute is dissolved in a solution.	C2 2.2			
12	Describe how to filter a mixture.	C2 2.4			
13	Use particle diagrams to identify mixtures.	C2 2.1			
14a	Compare the reactions of different metals with dilute acids.	C2 3.1			
14b, c	Identify different types of variable.	WS 1.1			
15	Calculate the amounts of metals in ores.	C2 3.5			
16	Identify displacement reactions from word equations.	C2 1.4			
17a	Explain the meaning of solubility.	C2 2.3			
17b	Interpret solubility graphs.	C2 2.3			
17c	Extract data from solubility graphs.	C2 2.3			
18	Explain why the concentration of carbon dioxide in the atmosphere did not change for many years.	C2 4.5			

P1.1 Charging up

A The particles that make up an atom are important to understand electricity.
Label the diagram with these keywords:

electron	proton	neutron

B Rubbing a balloon on a jumper leaves the balloon with an overall negative charge.

Complete the diagram to show where the charges are after this has happened.

C **a** Complete the sentences below to explain how objects can become charged.

Atoms are made of positive and negative particles. _____________________ separates charge. When the objects

are rubbed together, this moves _____________________ from one material to the other. The material that lost

_____________________ now has _____________________ charge overall. The other material has _____________________

charge.

b A series of objects have been charged with friction.
Complete the table to show which pairs of objects will attract, repel, or have no effect on each other.

Pair	1st object charge	2nd object charge	Attract, repel, or no effect?
A	positive	positive	
B	positive	negative	
C	negative	positive	
D	negative	negative	

c Charged objects have an electric field. Describe what is meant by an electric field.

____

____

D At a party, Luke touches a balloon against his hair and there is no reaction. After rubbing the balloon against his hair, his hair now sticks to the balloon. Explain why this happens.

____

____

What you need to remember

Atoms contain equal numbers of _____________________ (+) and _____________________ (−), so overall, atoms

have a _____________________ charge. When two objects are rubbed together, _____________________ moves

_____________________ from one object to the other, making one object _____________________ charged and the other

object positively charged. Charged objects are surrounded by an _____________________ _____________________. If two

charged objects have the same type of charge, they _____________________ each other. If two charged objects have

the opposite type of charge (one + and one −), they _____________________ each other. Thunderstorms generate very

strong electric fields, which cause _____________________ .

A a Define the term 'current'.

____

b Circle the name of the piece of equipment shown in the diagram, used to measure current.

ammeter   **voltmeter**   **newtonmeter**

B Complete the sentences below using these keywords to show how components affect the flow of electrons through a circuit.

opening	pushing	switch	cell	closing	switch	complete

Each circuit needs a power source, such as a battery or ___________________, to cause an electrical current

by ___________________ electrons around the circuit. They will only move if the circuit is ___________________.

___________________ a ___________________ completes a circuit, so that electrons can flow. ___________________ a

___________________ breaks a circuit, so that no electrons flow.

C A scientist decides to investigate the amount of electrical current needed to power different types of light bulb. She sets up a circuit for each lightbulb as shown.

a Complete the circuit to show how she can measure the current that the bulb needs.

b Tick the correct way in which the component in part **a** needs to be added to the circuit:

series

parallel

What you need to remember

Negatively charged ___________________ move when current flows in a metal. Current is the amount of charge

flowing per ___________________. We use an ___________________ to measure current in an electrical circuit. It must

be connected in ___________________ so that all of the current that flows through the component of interest also

flows through the ___________________. The unit of current is the ampere, often abbreviated to ___________________,

with the symbol ___________________. A component that can push charge around a circuit is a ___________________

or ___________________. A component that can make an object move is a ___________________. A device that can

break and complete a circuit is a ___________________. A circuit must be ___________________ for the charge to flow

around it.

P1.3 Potential difference

A Battery chargers need to be able to measure potential difference so they can sense when the battery has finished recharging.

   a Circle the correct unit for potential difference.

       ampere, A   **volt, V**   **ohm, Ω**   **watt, W**

   b Fill in the gaps to complete the sentences about potential difference.

      Potential difference (p.d.) means the 'push' provided by the ______________ (that is, the energy that it can provide). Another term commonly used for p.d. is ______________. The amount of potential difference that a ______________ provides is measured in ______________.

   c Describe how to measure the potential difference across a bulb.

      ____

      ____

   d Draw a circuit diagram for the equipment you would use in part **c**.

B An AA battery is labelled '1.2 V'. Describe what is meant by this rating.

____

C A mobile phone is powered by a battery. As the battery becomes flat, its potential difference decreases until it can no longer power the phone. Explain why a small potential difference cannot power the phone.

____

____

Hint: What is potential difference?

What you need to remember

Cells or batteries provide a push to charges in a circuit to make them move. This is called the ______________ ______________. We use a ______________ to measure potential difference. It must be connected in ______________ so that it is across the component we are interested in. Potential difference is measured in ______________, with symbol______________, so potential difference is often known as ______________. Cells and batteries are given a ______________, which is the potential difference across the cell / battery. A ______________ potential difference means that more ______________ is transferred to the components in a circuit than if the potential difference is ______________.

P1.4 Series and parallel

A Draw a line to match each property of a type of circuit with how it behaves.

| In a **series** circuit, **current** | across each component adds up to that across the cell. |

| In a **series** circuit, **potential difference** | through each branch adds up to that through the cell. |

| In a **parallel** circuit, **current** | across each branch is the same as that across the cell. |

| In a **parallel** circuit, **potential difference** | is the same everywhere. |

B Describe the difference between series and parallel circuits.

____

____

C The diagrams below show some simple series and parallel circuits, including some readings on ammeters and voltmeters. Write in the missing readings.

a   series potential difference   **b**   parallel current   **c**   parallel potential difference

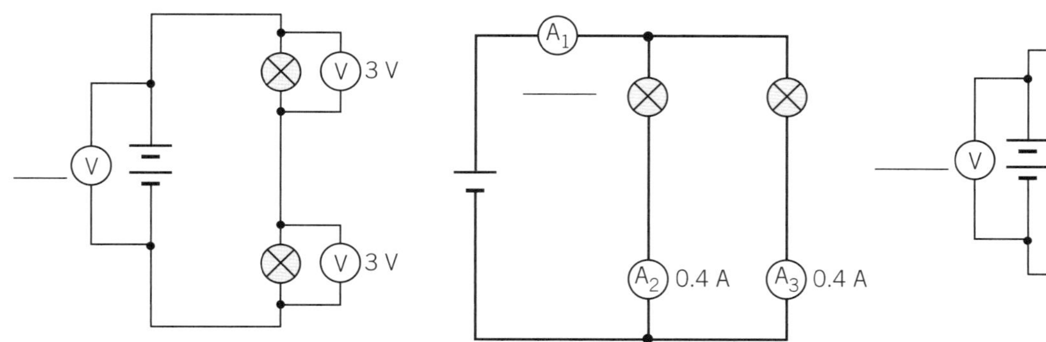

What you need to remember

There are two types of electrical circuits. _________________ circuits have all their components, including the cell, in one loop, so that there is only one path for the current to follow. As you add more components, the current gets _________________ . _________________ is the same everywhere. If you add up the _________________ across each component, the sum is equal to the _________________ across the cell.

_________________ circuits have their components in more than one loop, so there are at least two different paths for the current to follow. As you add more loops, each with their own components, the total current through the cell gets _________________ . If you add up the _________________ through each loop, the sum is equal to the _________________ through the cell. The _________________ _________________ across each loop is the same as that across the cell.

P1.5 Resistance

A One way engineers can find and fix problems with electrical circuits is to measure the resistance of the circuit. This helps them find broken components.

  a Describe what is meant by resistance.

____

____

  b Circle the electrical insulators below and underline the electrical conductors.

      copper   **plastic**   **gold**   **wood**   **glass**

  c Describe the difference between conductors and insulators in terms of resistance.

____

____

B Resistance can be calculated using the following formula:

$$\text{resistance } (\Omega) = \frac{\text{potential difference (V)}}{\text{current (A)}}$$

Calculate the resistance of the bulb in the circuit shown to the right. Give the unit.

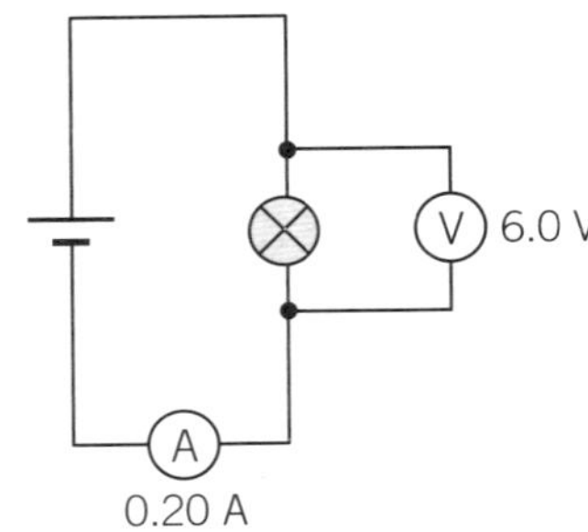

____

C a Rearrange the formula for resistance to give the formula for calculating current.

____

  b Calculate the current flowing through this cell.

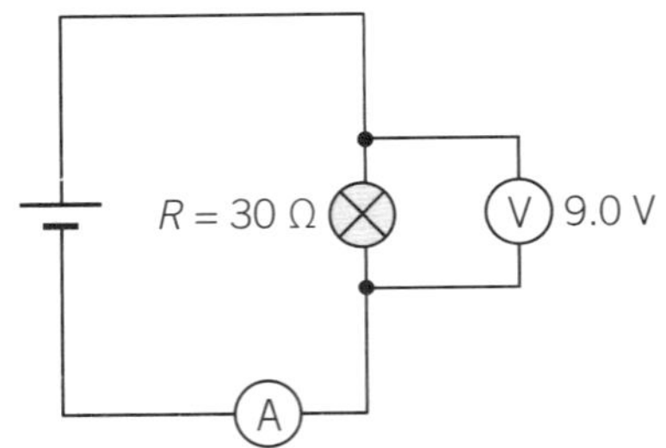

____

D The formula for resistance can also be rearranged to calculate potential difference:

$$\text{potential difference (V)} = \text{current (A)} \times \text{resistance } (\Omega)$$

Calculate the potential difference across this bulb.

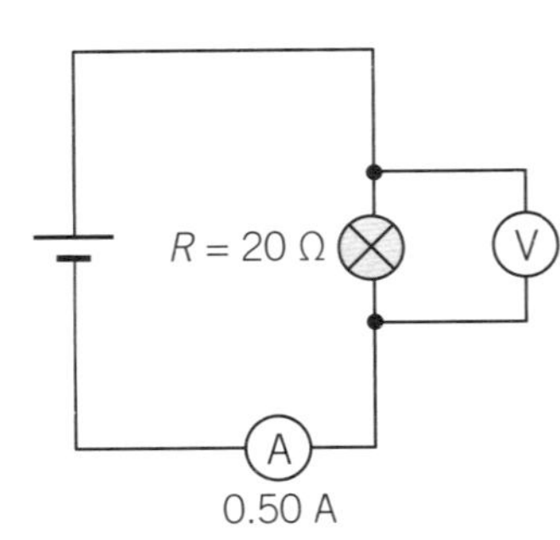

____

What you need to remember

_______________ is a measure of how much 'push' is required to get current through a component. Strictly, it

is defined and calculated as the _______________________________ across a component divided by the

_______________ through that component. It has the unit _______________ , symbol Ω. If 1 V applied across

a component causes 1 A to flow, we say that the component has a _______________ of 1 Ω. Insulators have very

_______________ resistance and conductors have very _______________ resistance.

P1.6 Magnets and magnetic fields

A **a** Complete this diagram to show the magnetic field lines around a bar magnet.

N   S

    b Complete this diagram to show the magnetic field lines around a bar magnet which is **stronger** than in part **a**.

N   S

B For each combination of magnets below, write whether they will attract, repel, or have no effect.

a	N	S	_______________________
b	S	N	_______________________
c	N	N	_______________________
d	S	S	_______________________

C Complete the diagram to show the Earth's magnetic field.

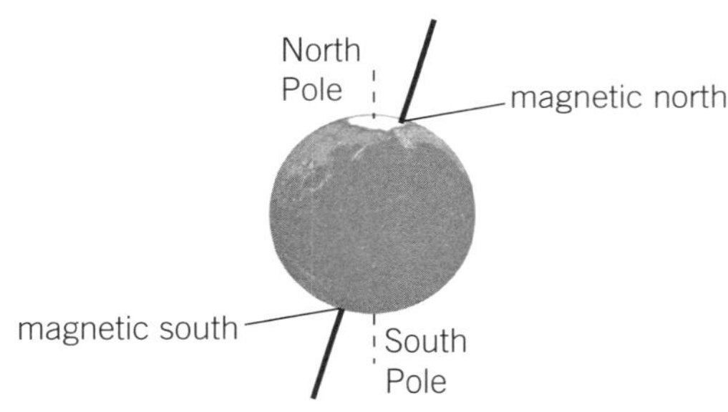

What you need to remember

Some materials are attracted to a magnet , or can be themselves turned into permanent ________________ .
These are called ________________ ________________ , and include the elements ________________ , nickel,
and cobalt, and some types of steel, which contains ________________ . Every magnet has a ________________
________________ and a ________________ ________________ . A compass is a magnet that is free to
rotate. It will spin until its ________________ seeking pole points to the ________________ north pole, which is
actually a magnetic ________________ pole. Two magnets will attract each other if they have ________________
poles nearest to each other, or repel if they have the ________________ poles nearest to each other. We say that
magnets are surrounded by a ________________ , which we can detect using a compass or
iron filings. We can represent this by drawing ________________ ________________ ________________ ,
which point from the north-seeking pole to the south-seeking pole, and show where a field is strong by drawing
________________ of them.

P1.7 Electromagnets

A a Give **one** difference between permanent magnets and electromagnets.

b Sketch an electromagnet and label its main features.

c Describe how to make an electromagnet.

B Describe **two** ways to make an electromagnet stronger.

C A group of students wants to investigate the effect current has on the strength of an electromagnet. They are provided with an electromagnet, an iron rod, a variable power supply, an ammeter and many steel paper clips.

a Identify the independent, dependent, and two control variables for their investigation.

independent variable _______________________________

dependent variable _______________________________

first control variable _______________________________

second control variable _______________________________

b Describe how they should carry out the experiment, including how they should use the equipment and what results they should collect.

What you need to remember

An electromagnet consists of a _______________ of wire with many _______________, wrapped around a _______________. To make an electromagnet stronger, create _______________ on the coil, use a larger _______________, or use a _______________ in the core that is easy to _______________, such as iron. Electromagnets can be more useful than _______________ magnets because they can be _______________ or made far stronger. The magnetic _______________ around an electromagnet is very similar to that around a _______________ magnet.

P1.8 Using electromagnets

A One use for electromagnets is in making an electric motor rotate.
Label the diagram to show the main parts of a motor.

B The statements below can be reordered to correctly describe
how a simple motor works.

Write the order of the statements that gives the best
description.

Correct order ☐ ☐ ☐ ☐

1 Current flows in coil.

2 Connect coil to a battery.

3 Coil becomes an electromagnet.

4 Forces between the coil and the permanent magnet make the coil spin.

C Describe two uses of electromagnets other than in a motor.

____

____

D A relay is a component which uses a small current to control a device which uses a lot of power.
Label this circuit diagram using the following phrases:

relay	**low current control**	**high power device**

What you need to remember

One device that uses electromagnetism to make things move is a ___________________ . A simple one consists of a

___________________ of wire near a pair of ___________________ ___________________ . Once connected to a power

supply, a ___________________ flows. The coil acts as an ___________________ , so there are forces between the coil

and the permanent magnets, making the coil ___________________ . Another device is a ___________________ .

This uses a small current in one circuit to control a ___________________ current in another ___________________ .

When the user closes a switch, a coil becomes an ___________________ . Two pieces of iron inside the coil are

___________________ . They attract each other, touching and ___________________ the other circuit. This is useful for

___________________ , so that the person controlling the current can be kept further away from large currents or

potential differences.

P2 Chapter 1 Pinchpoint 

Pinchpoint question

Answer the question below, then do the follow-up activity **with the same letter** as the answer you picked.

Choose the statement about electrical circuits that is correct.

A   Potential difference (voltage) measures the 'push' from the cell. Current is the amount of charge flowing per second.

B   In a series circuit with a bulb, there is less current after the bulb because current is used up in it.

C   Potential difference (voltage) is part of current: if there is no current then there can be no potential difference.

D   A cell always provides the same amount of current no matter which components are attached to it.

Follow-up activities

A The diagram in activity **D** shows a simple circuit in which a cell lights a bulb.

   a   Recall the formula used to calculate current. A cell rated at 3.0 V is used with a bulb rated at 3.0 V with a resistance of 10 Ω. Calculate the current that flows.

   b   The cell is changed to one rated at 1.5 V (no other components are changed). Predict any changes to the current in the circuit and the brightness of the bulb.

   c   The cell is changed to one rated at 12 V (no other components are changed). Predict what will happen to the current in the circuit and the bulb.

   Hint: What does the bulb rating mean? See P2 1.3 Potential difference for help.

B **a**   Give the definition of current.

   b   Charge cannot be created or destroyed. If one unit of charge enters a bulb, how much charge must leave the bulb?

   c   If one ampere of current enters a bulb, how much current must leave the bulb?

   d   A similar argument applies to parallel circuits, such as that shown in the diagram.

   The total current in the branches must equal that through the supply.

   Suggest how much current is at the point marked on the diagram.

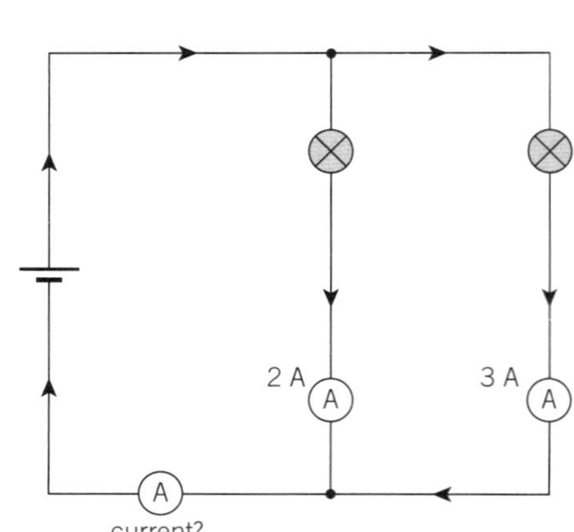

   Hint: What happens to the current in a series circuit? See P2 1.4 Series and parallel for help.

C A cell has a potential difference across its terminals even when it is not connected to a circuit, so no current is flowing. The photograph shows a cell with its battery rating.

a What does the term 'battery rating' mean?

____

b Complete the sentences using these keywords:

no current	**potential difference**	**a current**	**open**

When a switch is __________________ so the circuit has a break, there is __________________. If

__________________ is through a component, then there must be a __________________ __________________

across that component, but there is still a potential difference across a cell when it is not connected in a circuit.

Hint: Which component can create a potential difference? See P2 1.3 Potential difference for help.

c Draw a line to match each picture with its correct scientific name. You can use options more than once.

i 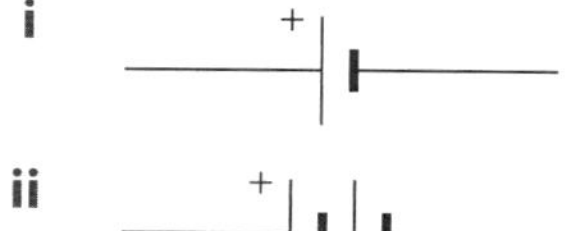

ii

cell

iii 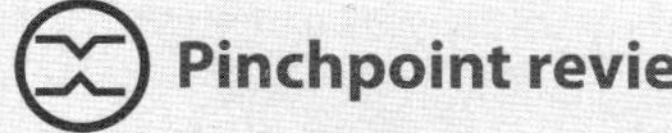

battery of cells

D a Complete the sentences using these keywords:

current	**connected**	**potential difference**	**resistance**

Cells and batteries are labelled with a rating which shows the __________________ __________________ they

provide at all times, whether they are __________________ to a circuit or not. The __________________ that

flows through the external circuit depends on the __________________ of that circuit.

b The circuit shown can be used to investigate the relationship between the resistance of the circuit and the current that the cell pushes through it.

Recall the formula used to calculate current. A cell rated at 3.0 V is used.

Calculate the current that flows if the circuit has a resistance of:

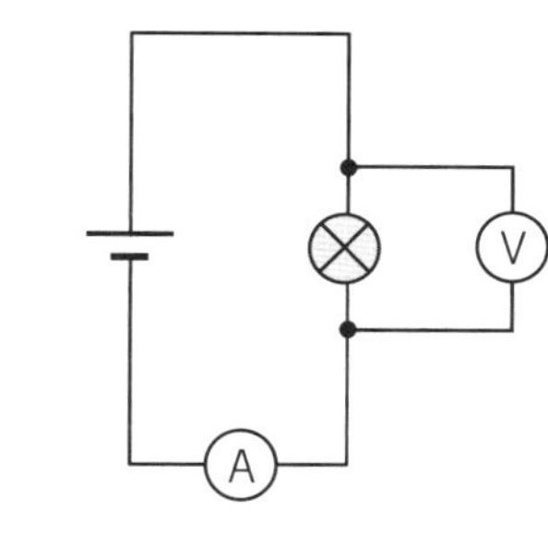

i 1.0 Ω

____

ii 20 Ω

____

Hint: Which factors affect the current through a component? See P2 1.5 Resistance for help.

(X) **Pinchpoint review**

Now look back at the question – do you think you chose the right letter?
Turn to the Answers page to find out.

P2.1 Food and fuels

A **a** Write **two** statements comparing the energy values of the foods shown in the table.

Food	Energy per 100 g (kJ)
Apple	200
Peas	250
Chips	1000
Chocolate	1500

b A teenager eats 100 g of chocolate during his 20-minute walk home from school. Walking requires 13 kJ of energy per minute.
Calculate whether the chocolate eaten has supplied more or less energy than that needed for his walk.

B **a** Draw a line to match each activity with its energy requirement.

| Sleeping | 300 kJ per hour |

| Running | 600 kJ per hour |

| Working | 3600 kJ per hour |

b Explain the different energy requirements for sleeping and running.

C A female office worker consumes 9000 kJ of energy in her food in one day.
A female swimmer training for the Olympic Games consumes 16 000 kJ in one day.
Both are eating healthy diets.
Explain why their food intake needs to be different.

What you need to remember

Different foods are stores of different amounts of ______________. The unit for this is the ______________,
and 1000 of them make up one ______________. A healthy diet will take in just as much energy as is needed
for the person's ______________. The body will convert extra food into ______________ to store unused
energy.

P2.2 Energy adds up

A Name **two** transfers from the chemical store of a torch to the thermal store of the surroundings.

____

B For each situation below, name the energy stores involved.

   a Coal is burnt to heat some water.

____

   b Georgia starts her petrol car. It begins to move and then speeds up.

____

   c Charlie freewheels down a hill on his scooter.

____

C For each situation in activity **B**, explain how each energy store empties and fills.

   a ____

____

____

   b ____

____

____

   c ____

____

____

What you need to remember

Energy cannot be created or destroyed – this is the law of _________________ of _________________.

Energy transfers between _______________________________, with some stores emptying as others

_________________, so that the _________________ amount of energy remains the same. Chemicals, such as fuels

or food and the oxygen needed to combust or respire, have a _________________ _________________ associated

with them. Hot objects are associated with a _________________ store; moving ones with a _________________

store; objects above the surface of the Earth with a _________________ _________________ store; stretched or

compressed ones an _________________ store.

In many processes, energy is not only transferred to the store you want it in, but also 'lost' to a store which is not

useful. We say that that energy has _________________. For instance, when heating the water in a kettle, energy

might be transferred not only to the _________________ store associated with the water, but also heat the room,

filling the room's _________________ store, which is not useful.

P2.3 Energy and temperature

A When an object is heated, both its temperature and the energy in its thermal store increase.
Fill in the gaps to complete the sentences below about temperature and energy.

increases	J	°C	move / vibrate	thermal	increases	stays the same

Temperature is measured in ________________. As the mass of an object changes, its temperature

________________. As temperature ________________, the particles that make up the object

________________ more.

Energy is measured in ________________. As the mass of the object increases, the amount of energy in its

________________ store ________________.

B a Ice is composed of water particles. Describe what happens to them when you heat it.

____

b Describe what happens to the particles in liquid water when you heat it.

____

C a Katie places a bowl of hot soup on the table in the kitchen. Describe what happens to the temperatures of the
soup and the air in the room.

____

____

b Katie places the cold, empty bowl in warm water in the sink. The temperatures of the bowl and water will
change until they reach equilibrium.
Explain what is meant by equilibrium.

____

____

What you need to remember

We describe how hot or cold something is as its ________________. We measure this using a

________________, using the unit of ________________ ________________, symbol ________________.

If a hotter object is put in contact with a colder one, the hot one will heat the cold one until they reach

the ________________ temperature and are in ________________, that is until ________________

________________ energy is transferred between their ________________ stores.

The energy that you need to raise the temperature of a material depends on the ________________ of material

and the ________________ of material, as well as on how much you want to raise the temperature.

P2.4 Energy transfer: particles

A Circle the correct keywords to complete these sentences.

A thermal insulator can **increase / reduce** heat loss compared to conductors. Energy is transferred much **slower / faster** through an insulator. Gases are good **insulators / conductors** because their particles are far apart.

B a The photograph shows an example of conduction.
Explain how energy is transferred during conduction.

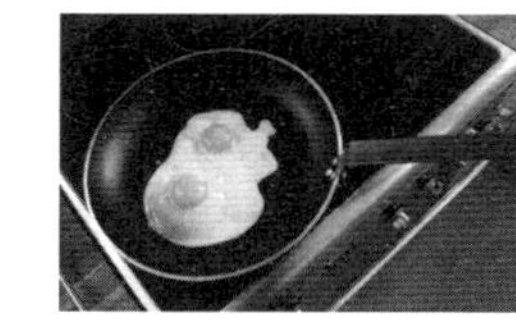

____

____

b The photograph shows an example of convection.
Explain how energy is transferred during convection.

____

____

C The diagram shows an experiment to measure which materials are the best conductors.

A timer is started when the ends of the rods are heated and stopped when the wax at the other end melts. The table shows the results for three materials.

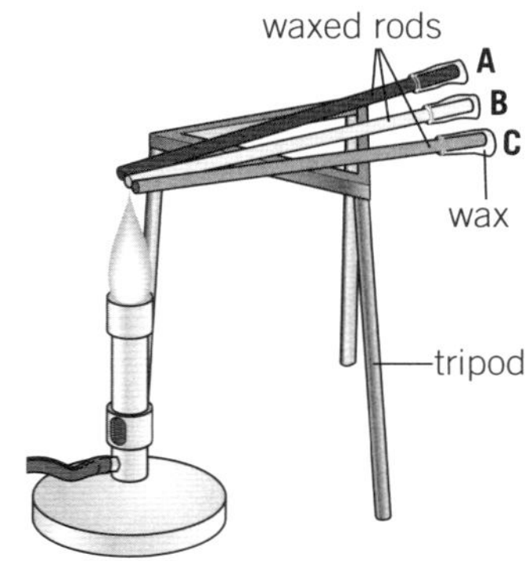

Rod	Material	Time taken (s)
A	glass	More than 1000
B	aluminium	80
C	copper	48

Write a conclusion, including which material is the best conductor.

____

____

____

____

What you need to remember

Energy can be transferred by heating – this transfers the energy from a _______________ store associated with a

_______________ object into the thermal store of a _______________ object. This can happen in three ways.

Particles in a hot material _______________. When they collide with their neighbours, making them vibrate,

we call that _______________. This process happens fastest with materials in a _______________ state and

slowest with materials in a _______________ state.

When you heat a liquid or gas, its particles move further apart so the fluid becomes less _______________.

The hotter, less _______________ fluid then rises, moving to a place which is colder. This process is called

_______________. The movement of the fluid from one place to another is called a _______________

_______________.

A a   Wet laundry can be dried quickly by hanging it outside on a sunny day. Explain why black clothes dry more quickly than white ones.

b   Water can be heated by the Sun using the device shown. The water gets hottest when matt black paint is used in the device, rather than shiny black or any other colour or finish.
Explain why.

B a   Draw a line to connect each method of energy transfer to its cause.

Conduction	Emission and absorption of infrared
Convection	Particles moving from a hotter place to a colder place
Radiation	In solids, particles vibrating and colliding with neighbours

b   Explain how energy is transferred by radiation using these keywords:

radiation   **emit**   **absorb**   **infrared**   **hot objects**   **heating**

c   Suggest one piece of evidence that radiation does not need particles to transfer energy.

Hint: What do you feel when you turn your face to the Sun?

What you need to remember

To transfer energy by _________________ or _________________ requires particles. However, these are not needed

for heating by _________________. Hot objects emit _________________ _________________,

sometimes known as thermal radiation or heat. This can be detected using a _________________

_________________ _________________, for instance to help firefighters find people in a smoke-filled

room. When objects _________________ this radiation, they warm up. It is a wave like light and can be

_________________, _________________, or _________________. Surfaces which have a _________________

colour and _________________ finish absorb infrared better than ones which are _________________ and

_________________. The best absorbers are also the best emitters, so _________________, _________________

surfaces emit the most infrared.

P2.6 Energy resources

A **a** List three renewable and three non-renewable resources in the table below.

Renewable	Non-renewable

b Describe the difference between renewable and non-renewable energy resources.

____

____

B **a** Name an energy resource that can be used in a thermal power station, other than coal.

____

b Write the order of these statements which gives the best description of how coal is used in a thermal power station to generate electricity for your home.

Correct order ☐ ☐ ☐ ☐ ☐ ☐

1 steam, which drives a…

2 Coal is burnt, which heats…

3 electrical current to power your home

4 turbine, which turns a…

5 water, to produce…

6 generator, which pushes…

C Compare the advantages of renewable and non-renewable energy resources.

Renewable ____

Non-renewable ____

P2.7 Energy and power

A Draw a line to match each quantity with its unit and its description

| Energy | W | Cannot be created or destroyed |

| Power | J | How quickly stores are emptied and filled |

B Choose the correct equation from this list to use in each of the questions below.

$$\text{power (W)} = \frac{\text{energy transferred (J)}}{\text{time taken to transfer (s)}}$$

energy (J) = power (W) × time (s)

power (W) = current (A) × potential difference (V)

energy (kW h) = power (kW) × time (h)

cost (p) = power (kW) × time (h) × cost per unit (p / kW h)

a A gas stove runs for 30 s and burns fuel of energy 45 000 J.
Calculate the power produced by the gas stove.

____

b **i** A hairdryer requires a current of 10 A and operates at 230 V.
Calculate the power produced by the hairdryer.

____

ii Calculate how much energy is transferred by the hairdryer in 30 s.

____

c An electrical radiator to heat a room is rated at 2 kW.

i Calculate the energy transferred when the radiator heats a room for four hours.

____

ii If the electricity company charges 15 p / kW h, calculate the cost of using the radiator.

____

What you need to remember

Power (in W) is defined and calculated as the _______________ transferred (J) divided by the

_______________ taken for that transfer (s). If one joule of _______________ is transferred in one second,

then the _______________ produced is one watt (W). In electrical circuits, power can be calculated as the

_______________ times the _______________. Electrical devices are often labelled with

a _______________. This shows the amount of power they can produce. The unit for this is

the _______________ with the symbol W, and 1000 of them make up one _______________, symbol kW. If you

use a device that produces one kilowatt of power for one hour, you will have transferred one _______________

_______________ of energy, symbol kW h.

P2.8 Work, energy, and machines

A Simple machines include devices like levers. Fill in the gaps to complete these sentences about levers. You may use a word more than once.

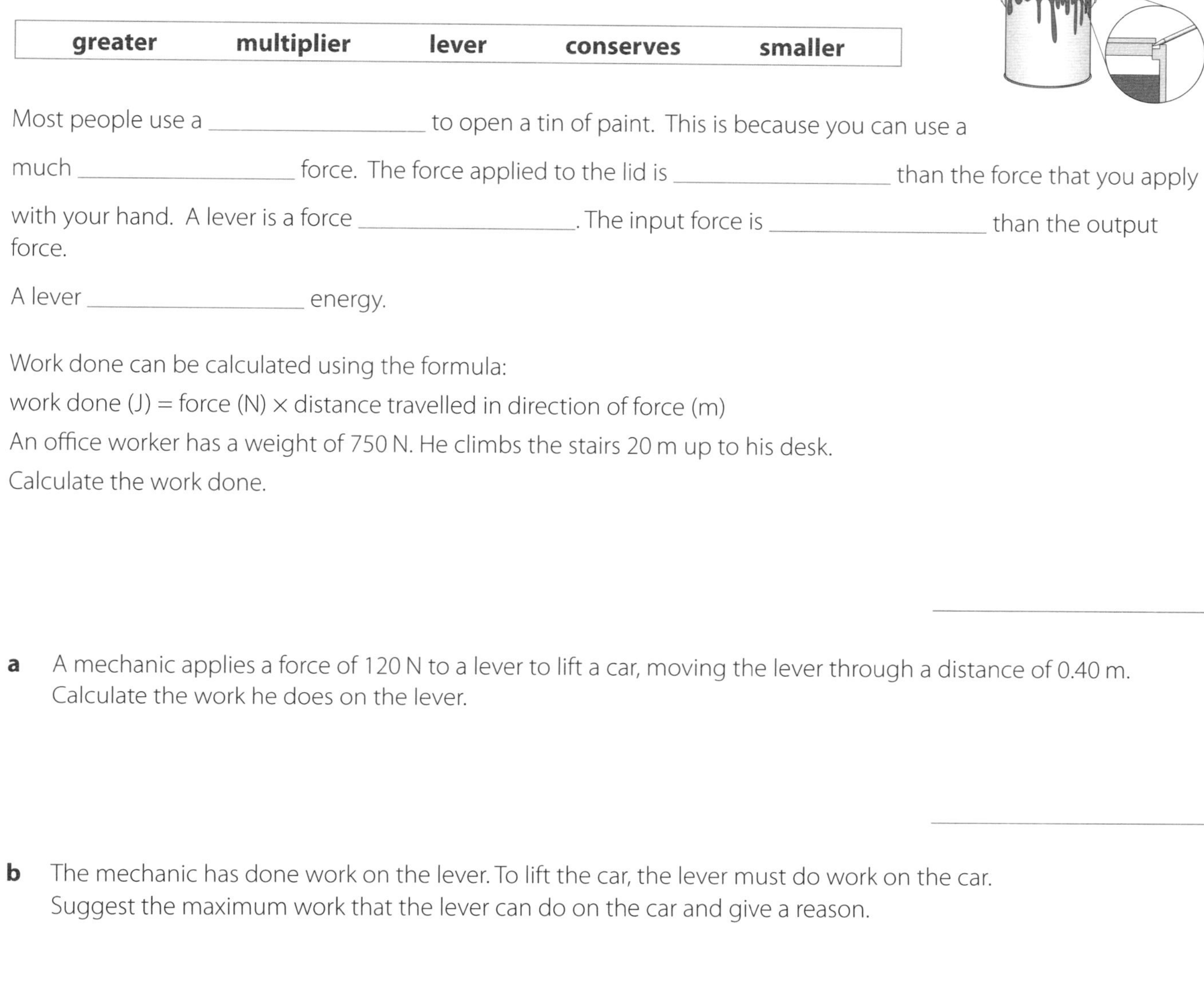

greater	multiplier	lever	conserves	smaller

Most people use a ______________ to open a tin of paint. This is because you can use a

much ______________ force. The force applied to the lid is ______________ than the force that you apply

with your hand. A lever is a force ______________. The input force is ______________ than the output
force.

A lever ______________ energy.

B Work done can be calculated using the formula:

work done (J) = force (N) × distance travelled in direction of force (m)

An office worker has a weight of 750 N. He climbs the stairs 20 m up to his desk.

Calculate the work done.

____

C **a** A mechanic applies a force of 120 N to a lever to lift a car, moving the lever through a distance of 0.40 m.
Calculate the work he does on the lever.

____

b The mechanic has done work on the lever. To lift the car, the lever must do work on the car.
Suggest the maximum work that the lever can do on the car and give a reason.

____

What you need to remember

When an object is moved by a force through a distance, we say that ______________ is done. It is a way

of transferring ______________, like heating. Work ______________ is defined and calculated as the

force (N) times the distance moved (m). It has the same unit as energy, the ______________, symbol J. If a

______________ of one newton moves an object one metre, then one joule of ______________ was done,

and one joule of ______________ was transferred between stores. Devices that reduce the force needed to

move an object, or increase the distance the object moves when you apply a force, are called ______________

______________. Examples include ______________, pulleys, and ______________.

Pinchpoint question

Answer the question below, then do the follow-up activity **with the same letter** as the answer you picked.

Nathan uses a gas camping stove to heat some soup.

Which statement about the energy stores is correct?

A The total energy lost from stores that empty must be equal to the total energy gained by stores that fill.

B The thermal store of the fuel empties.

C The total energy lost from stores that empty must be greater than the total energy gained by stores that fill.

D The chemical store of the soup empties.

Follow-up activities

A a Complete the following sentences using the keywords below.

thermal store	dissipation	minimum	thermal store	lots

Many processes involve ________________, in which some energy is transferred to the ________________ of the surroundings. We would like all of the energy input during a process to be transferred only to the store that we want; however, this is often impossible. An efficient process keeps the energy 'lost' by dissipation to a ________________. An inefficient process means that ________________ of energy is transferred to the ________________ ________________ of the surroundings rather than to the store that we want.

b i How could you make the process of heating the soup more efficient?

____

ii Explain why your suggestion(s) for part **i** would burn less fuel.

____

____

Hint: What is dissipation? See P2 2.2 Energy adds up for help.

B When an energy store changes, there is at least one quantity associated with it that we can observe change.

a Draw a line to match each store with its example process and an observation we could see when that store changes. Circle whether each store has filled or emptied.

Energy store	Example process	Observed change	Store emptied or filled?
Thermal	Burning petrol in a car engine	Speed of bike increases	Emptied / filled
Chemical	Accelerating bike	Length of elastic band increases	Emptied / filled
Kinetic	Cooling soup	Temperature decreases	Emptied / filled
Gravitational potential	Stretching elastic band	Height of bag increases	Emptied / filled
Elastic	Lifting bag of shopping	Mass of fuel reduced	Emptied / filled

b What happens to the fuel as it is burnt to heat the soup?

____

c Which energy store associated with the fuel changes when we burn it?

____

Hint: What is the chemical energy store? See P2 2.2 Energy adds up for help.

C a What happens to your body temperature when you exercise hard?

____

b When the oven is switched on in the kitchen, what happens to the temperature of the room?

____

c Most computers have fans inside them. Suggest why the fans are needed.

____

d Many processes involve not only the desired change (such as driving a car along a road) but also some wasted effort, an unwanted side-effect of the change (such as the tyres and road heating up). This effect is called dissipation.

Which **store** of the surroundings fills during dissipation?

____

e Fill in the gaps to complete this statement about the law of conservation of energy.

Energy cannot just ______________, and you cannot end up with ______________ than you had at the start.

Energy cannot be ______________ or ______________, only ______________.

Hint: What is dissipation? See P2 2.2 Energy adds up for help.

D a Complete the table to show which store is involved and what is observed when it fills or empties.

Object	Store	Observed change when store fills	Observed change when store empties
Soup	Thermal	Temperature increases	
Fuel	Chemical		Mass decreases
Lift		Height increases	Height decreases
Car	Kinetic		

b Circle the correct term in the box below to show how the soup in the Pinchpoint question was changed.

chemically   **thermally**

Hint: Which energy stores are there? See P2 2.2 Energy adds up for help.

Pinchpoint review
Now look back at the question – do you think you chose the right letter?
Turn to the Answers page to find out.

P3.1 Speed

A Use the keywords to complete the sentences below about motion.

compared	stationary	25	relative

When we say that a car has a speed of 30 km/h, we are comparing it with something we think of as

_____________________, such as the ground. _____________________ motion means describing how one thing moves

_____________________ to another. For instance, if a car travelling at 30 km/h is overtaking someone walking at

5 km/h, we would say that the car has a speed of _____________________ km/h relative to the person walking.

B Average speed can be calculated by the formula:

$$\text{average speed (m/s)} = \frac{\text{total distance travelled (m)}}{\text{total time taken (s)}} \quad \text{OR average speed (km/h)} = \frac{\text{total distance travelled (km)}}{\text{total time taken (h)}}$$

a A motorbike drives a distance of 500 m in a time of 30 s.

Calculate its average speed.

____

b The formula for speed can be used with different units, as shown above.

A train travels 160 km in 2.0 h. Calculate the average speed.

____

c The formula for speed can be rearranged to calculate the distance travelled:

distance travelled (m) = speed (m/s) × time taken (s)

The sound of thunder travels at 330 m/s. Calculate how far it will travel in 12 s.

____

C The diagram shows two cars with different speeds and directions.

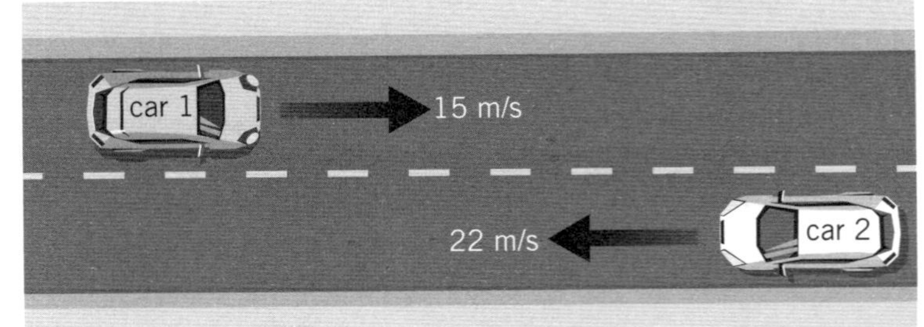

a What is relative speed?

____

____

b What is the **relative** speed of the two cars?

____

What you need to remember

The rate at which something moves is called _____________________. It is defined and calculated as the

_____________________ travelled (m) divided by the time taken (s). It has the unit _____________________

_____________________, symbol m/s. If something moves a _____________________ of

one metre in one second, then it has a _____________________ of one m/s. We can distinguish between the

_____________________ that something has just at one moment – its _____________________ _____________________ – and

its _____________________ _____________________ over a whole journey. When two objects are moving, we can talk

about their _____________________ _____________________. For instance, two people walking at the same speed of

1.5 m/s would have a _____________________ _____________________ of zero if they are walking in the same direction,

but a _____________________ _____________________ of 3 m/s if they are walking in opposite directions.

P3.2 Motion graphs

A For each graph, describe simply the object's movement.

a

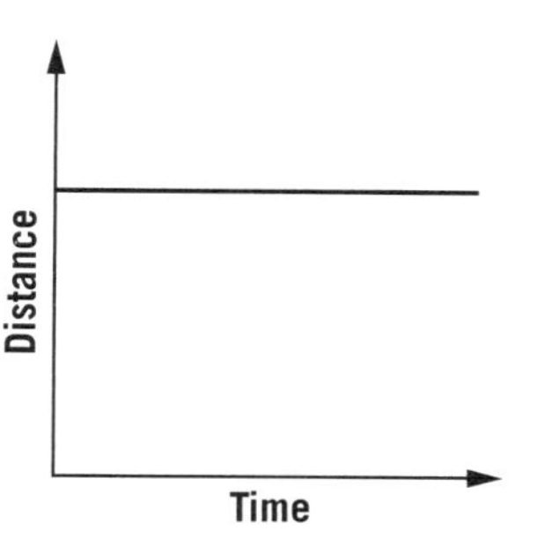

b

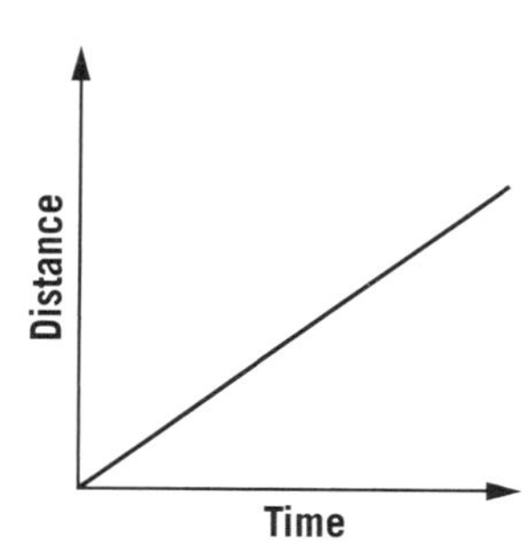

c

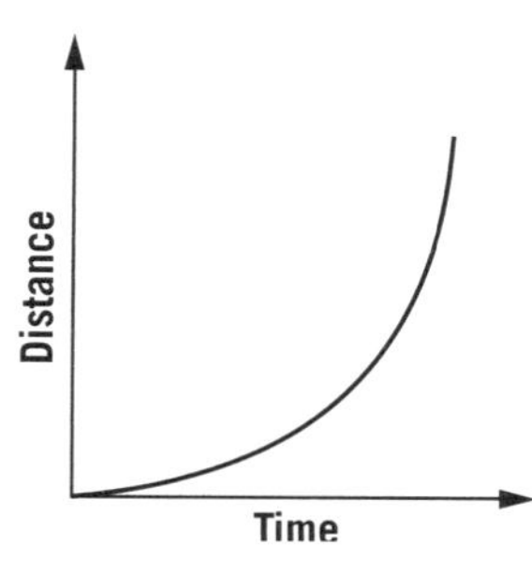

B a The table shows data for a journey by car. Plot the data on a distance–time graph.

Time (s)	Distance (m)
0	0
10	0
20	12
30	24
40	24
50	24
60	24
70	48
80	72

b Describe the journey shown in your graph.

c Calculate the speed of the car between 10 s and 30 s.

What you need to remember

One way to describe a journey is to plot a ________________–________________ graph. If an object is not

moving, the graph stays ________________. The slope of the graph shows its ________________. If the object's

speed is changing, we say it is ________________.

A When you pump up a tyre with a bicycle pump you change the pressure of the air in the pump.

   a Sketch the arrangement of air particles in the bike pump after being compressed.

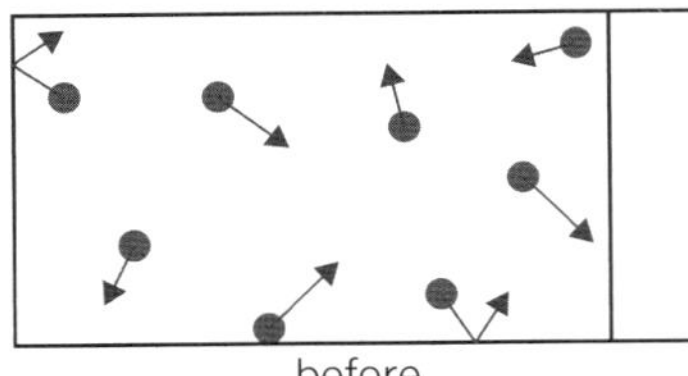

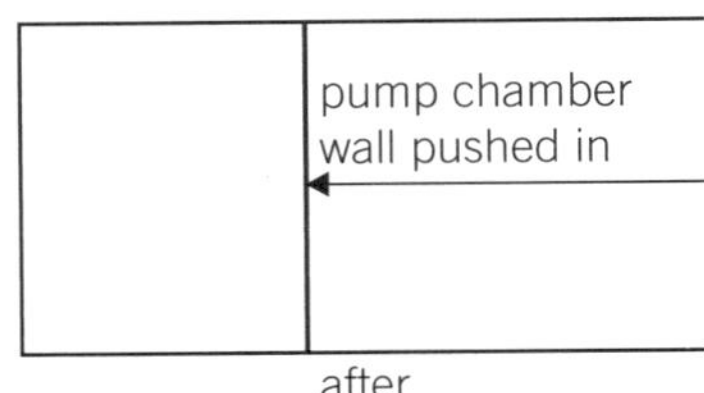

   b Explain why the cylinder of a bike bump must be made of a strong, rigid material.

____

B Draw a line to match each factor with its effect on pressure and the reason.

| An increase in **temperature** | causes an **increase** in pressure | because the particles have **further** to travel so collide **less** often. |

| An increase in **volume** | causes a **decrease** in pressure | because the gas particles move **faster**, so collide **harder** and **more** often. |

C The photograph shows a pilot dressed to cope with the effects of being at high altitude.

   a Explain what causes atmospheric pressure.

____

____

____

   b Describe how atmospheric pressure changes with height and **one** effect this has on people.

____

____

What you need to remember

Gases, such as air, exert _____________________________ because of collisions between the gas

_____________, and between them and any surface they touch. When we squeeze a gas it is

_____________, which increases its pressure and its _____________. The air around us is at

_____________________________, which causes a force pushing in on our skin. This is balanced by the

pressure from inside our bodies. As you go higher, such as up a mountain, _____________

_____________ gets _____________.

P3.4 Pressure in liquids

A a For each object, circle whether you expect it to float or sink when placed in water.

    boat: **sink**   **float**

    pebble: **sink**   **float**

    basketball: **sink**   **float**

b Describe why some objects float and others sink.

Float ____

Sink ____

B Describe how liquid pressure changes with depth as a submersible probe sinks deeper in the ocean.

____

____

C Complete each diagram to show the forces on each object.

a A boat floating.

b An anchor about to sink.

What you need to remember

Liquids, such as water, exert _____________________ because of how the particles push against each other and anything they touch. There is a difference in _____________________ between the top and bottom of an object which is in water. This causes a force called _____________________. This is what causes people to float when they are swimming. When liquids are squeezed their _____________________ hardly changes at all: they are

_____________________.

P3.5 Pressure on solids

A Pressure can be calculated by the formula:

$$\text{pressure (N/m}^2) = \frac{\text{force (N)}}{\text{surface area (m}^2)}$$

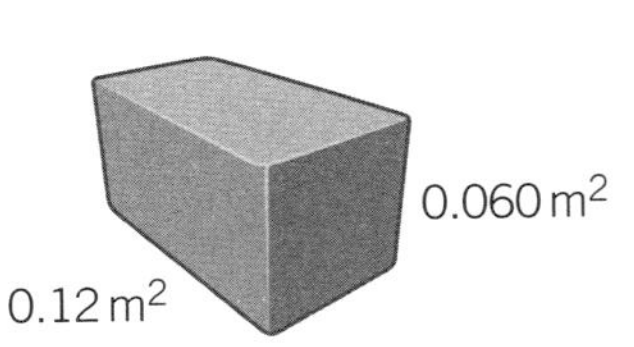
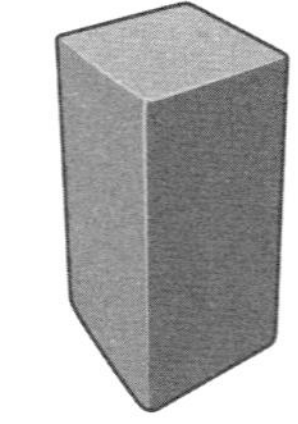

a The diagram shows a child's foam block lying flat and standing on its end. The block weighs 20 N.
Calculate the pressure on each face of the block.

Lying flat: ____

On its end: ____

b Three blocks are stacked on top of each other, lying flat.
Calculate how much pressure is on the floor under the bottom block.

____

B **a** A box of cereal has a weight of 5.0 N, and an area of 0.010 m².
Calculate the pressure it exerts on its kitchen shelf.

____

b A garden shed and its foundations have a weight of 2500 N, and an area of 6.4 m².
Calculate the pressure between it and the ground.

____

C Some historic homes have floors made of wood that need to be protected from damage.
Use the idea of pressure to explain why narrow 'stiletto' heels might damage the floor, but the same person wearing shoes with a wider heel is unlikely to.

____

____

____

What you need to remember

_________________ is defined and calculated as the _________________ applied (N) divided by the

_________________ over which it is applied (m²). It has the unit _________________

_________________, symbol N/m². If a _________________ of one

newton is applied over one square metre, there is a _________________ of 1 N/m². A _________________ force

exerted on a _________________ area exerts a large _________________ on the surface, whereas the same force

spread over a _________________ area will result in _________________ pressure on the surface.

P3.6 Turning forces

A The photograph shows a ruler balanced with two apples. This situation obeys the law of moments.

Give the law of moments using these keywords.

| equilibrium | sum | clockwise |
| anticlockwise | moments | |

____

____

B Moment of force can be calculated by the formula:

moment (N m) = force (N) × perpendicular distance from the pivot (m)

a Callum applies 5.0 N of force to a door handle 0.50 m from the hinges to open the door. Calculate the moment of force.

____

b Molly holds a 50 N bag of groceries. Her forearm is 0.40 m from her elbow to her hand. Calculate the moment of force of the groceries about her elbow.

____

c Tightrope walkers must keep their centre of gravity directly above the rope. Explain why this is important.

____

____

____

What you need to remember

When we apply a force to a door that can ______________ on its hinges, there is a turning effect called a ______________. The ______________ of a force is defined and calculated as the ______________ applied (N) times the perpendicular ______________ to the ______________ (m). It has the unit ______________ ______________, symbol (N m). If a ______________ of one newton is applied one metre from a ______________, then there is a ______________ of one N m. If the sum of the moments in a clockwise direction are equal to the sum of the moments in the anticlockwise direction, the object is in ______________ and it will not start turning. This is the ______________ ______________. The centre of gravity is the point through which the ______________ appears to act. If the centre of gravity of an object is directly ______________ or ______________ the pivot there will be ______________ moment making it turn. If it is to the side of the pivot, the object will start to ______________ due to the moment.

Pinchpoint question

Answer the question below, then do the follow-up activity **with the same letter** as the answer you picked.

Daisy is riding a bicycle down a straight road. This is the distance–time graph for her journey.

Choose the option below that correctly describes her motion at the specified time.

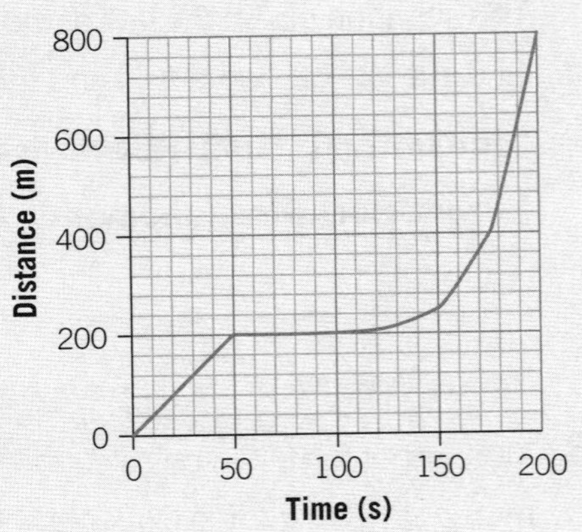

	Time range	Motion
A	From 0 s to 50 s	Daisy is stationary
B	From 100 s to 200 s	Daisy is accelerating and has constant speed
C	From 0s to 200 s	Daisy has an average speed of 4.0 m/s
D	From 0 s to 50 s	Daisy is accelerating

Follow-up activities

A a Complete the sentences using these keywords.

no	distance travelled	time taken	zero	horizontal (level)

Speed is ___________________ ___________________ divided by the ___________________ ___________________.

If someone travels ___________________ distance because they are stationary, their speed is ___________________.

On a distance–time graph, no change in distance means that the graph will not go higher or lower, it will stay

___________________.

b On these axes, sketch a graph to show Daisy:

i stationary

ii moving with increasing speed

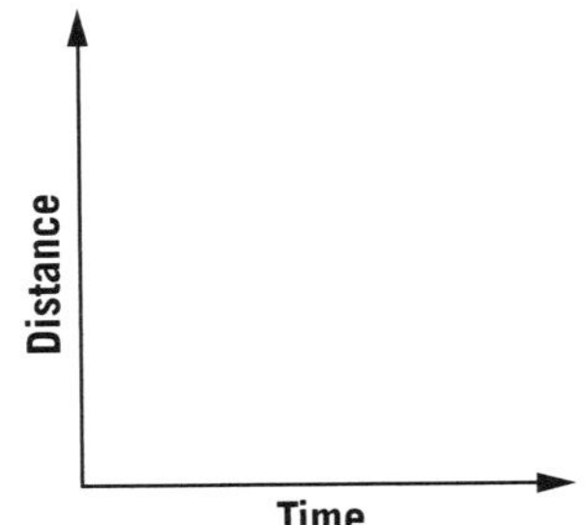

Hint: What does the slope of a distance–time graph tell us? See P2 3.2 Motion graphs for help.

B a Complete the sentences using these keywords.

steep	curve	slope	changing	shallow	speed

The ___________________ of a distance–time graph shows the ___________________ of the object. A fast object

is shown with a ___________________ line, a slow one with a ___________________ line. If Daisy is cycling in

a straight line and accelerating, then her speed must be ___________________, so the slope of the line must

___________________ upwards.

b On these axes, sketch a graph to show Daisy moving with:

    i **slow** constant speed   **ii** **fast** constant speed   **iii** increasing speed

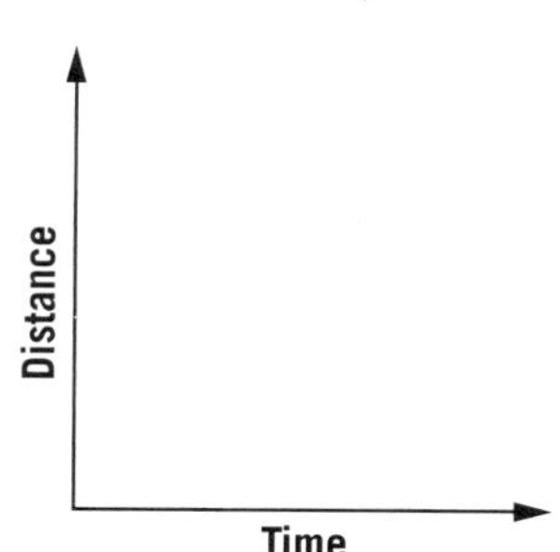
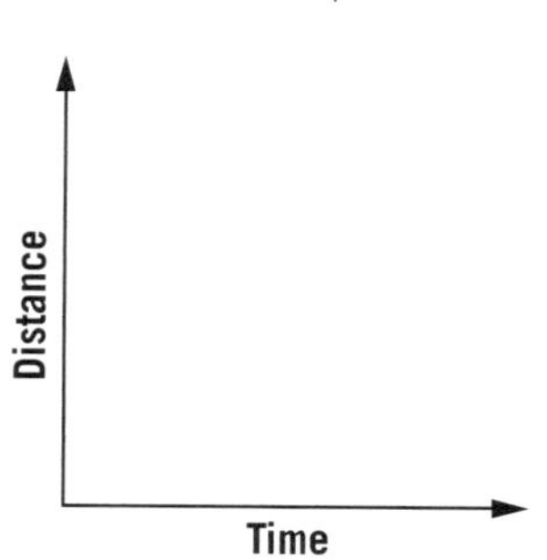
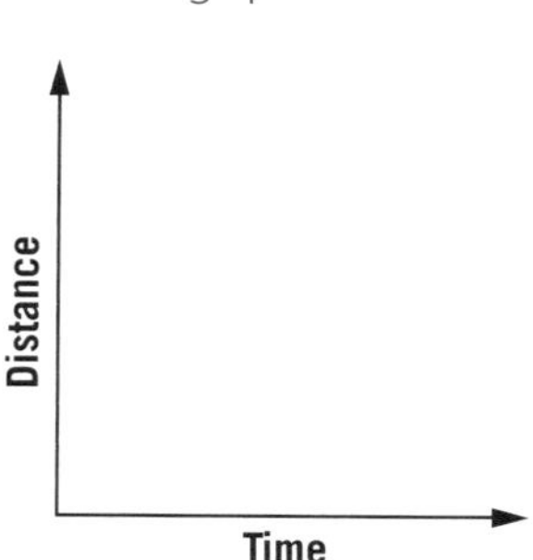

Hint: What does the slope of a distance–time graph tell us? See P2 3.2 Motion graphs for help.

C Balanced forces mean that an object is in equilibrium and its speed is not changing.
Circle the part of Daisy's journey that matches each force diagram.

a

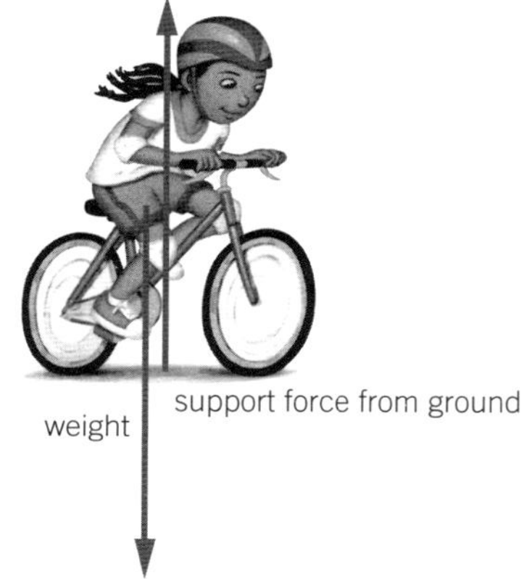

b

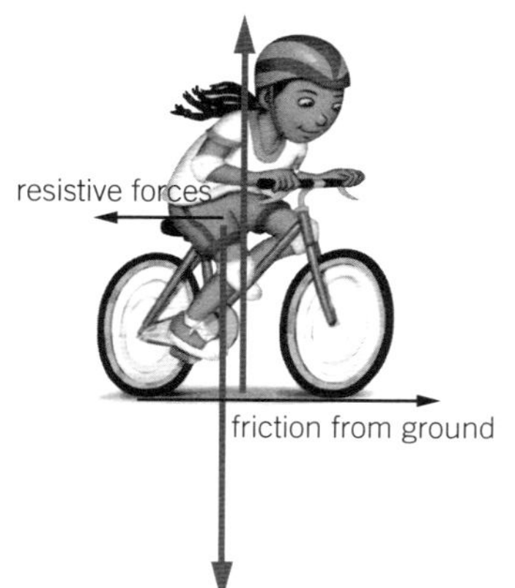

c

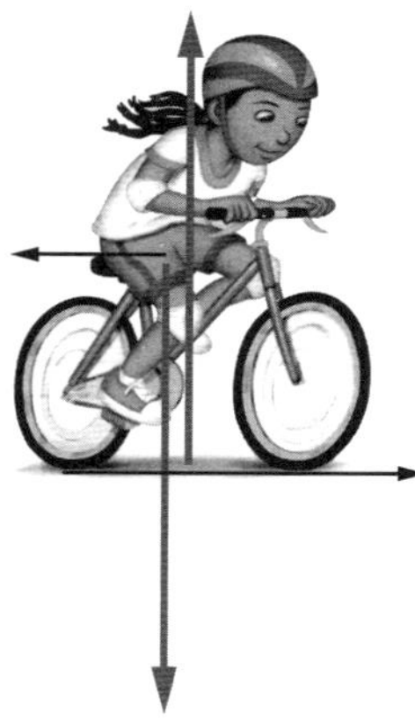

    0 s to 50 s / 50 s to 100 s /   **0 s to 50 s / 50 s to 100 s /**   **0 s to 50 s / 50 s to 100 s /**
          100 s to 200 s   **100 s to 200 s**   **100 s to 200 s**

Hint: How do unbalanced forces affect an object? See P1 1.5 Balanced and unbalanced for help.

D **a** Circle the correct **bold** word in each sentence.

To start moving, **unbalanced / balanced** forces are needed to cause Daisy to accelerate. Acceleration means her speed must **change / stay constant**. Once Daisy is moving, the acceleration can drop to zero while she continues moving with **increasing / constant** speed. On a distance–time graph, a straight line with constant slope shows **increasing / constant** speed, which means that she **can / cannot** be accelerating.

Hint: What is acceleration? See P2 3.2 Motion graphs for help.

b Sketch a graph on the axes given to show different parts of a journey in car, where the car is:

    i   accelerating
    ii   moving at a constant speed
    iii stationary

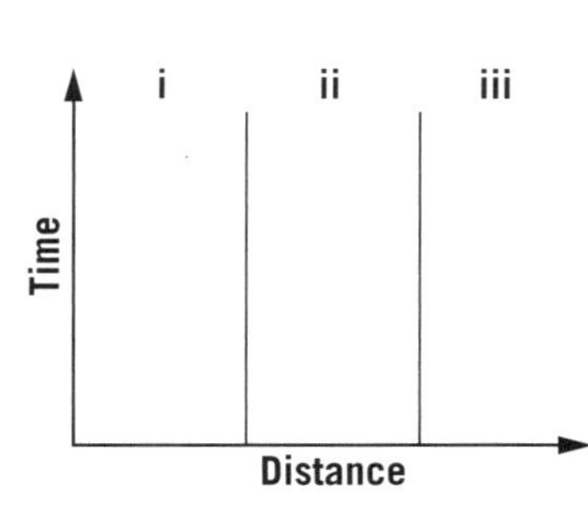

c Which of the sections of your graph show balanced forces? _____________

 **Pinchpoint review**

Now look back at the question – do you think you chose the right letter?
Turn to the Answers page to find out.

P2 Revision questions

1 **a** Tick **two** uses of electromagnets. *(2 marks)*

   1 Separating plastics for recycling ☐

   2 Motor for electrically powered train ☐

   3 Separating metal for recycling ☐

   4 Lighting with an electric light bulb ☐

b For one of the uses you chose, describe how the electromagnet is used. *(2 marks)*

____

____

2 A toaster works mainly through radiation.

a Draw a line to match each stage of the process to the object. *(3 marks)*

Transmitted through	the heating element
Absorbed by	the toast
Emitted from	the air

b Temperature and the amount of energy in the thermal store are related to each other.

Give the units for temperature and energy.

Temperature _________________ *(2 marks)*

Energy _________________

3 **a** Draw a circuit diagram to show how to measure the potential difference across a bulb powered by a cell. *(2 marks)*

b Name the component you have drawn to measure potential difference. *(1 mark)*

____

4 **a** Circle the quantity that flows in an electric circuit. *(1 mark)*

resistance   potential difference

charge   components

b When a potential difference of 12 V is applied across a bulb, a current of 4.0 A flows through it. Calculate the resistance of the bulb. Give the unit.

The formula for calculating resistance is:

$$\text{resistance} = \frac{\text{potential difference}}{\text{current}}$$ *(2 marks)*

____

c The wires leading into the bulb are wrapped with plastic with a resistance of 1 000 000 Ω. Circle whether the plastic wrapping and the filament wire are conductors or insulators.

filament wire: **conductor / insulator**   *(1 mark)*

plastic wrapping: **conductor / insulator**   *(1 mark)*

5 Rashida carries out an investigation of current in series and parallel circuits.

a She takes the measurements shown in **Figures 1** and **2**. Identify the differences in the way the current flows in the two types of circuit.

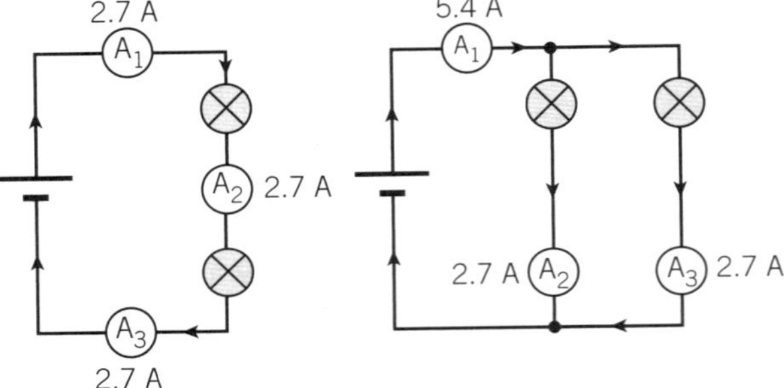

Figure 1   **Figure 2**

Series circuit: _________________ *(1 mark)*

Parallel circuit: _________________ *(1 mark)*

____

b Rashida sets up a new circuit. Predict the readings of potential difference she will find and fill them in on **Figure 3**. *(2 marks)*

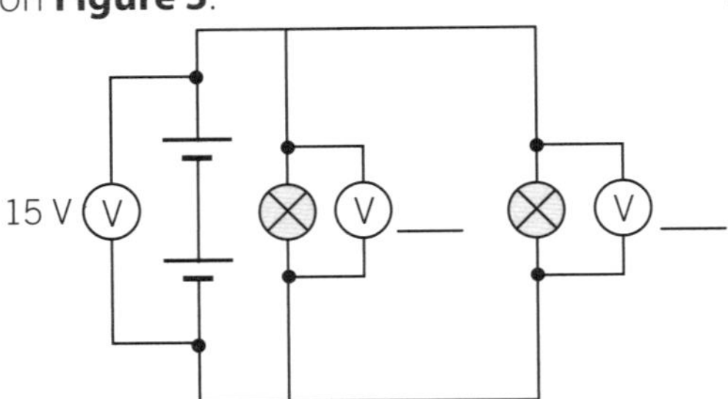

Figure 3

6 **a** In 2006, Hannah McKeand set the record for reaching the South Pole in the fastest time. In one hour she travelled approximately 3000 m pulling a sledge of food and equipment requiring a force of about 200 N.

Calculate the work she did in one hour and give the unit of work. The formula for calculating work done is:

work done = force × distance (*2 marks*)

b On her trip, McKeand needed to absorb 3000 kJ of energy from her food each hour. While working in an office, a typical adult would need to consume 400 kJ of energy in her food every hour.

Is McKeand's energy intake larger or smaller than that of an office worker? Suggest why. (*2 marks*)

7 In 2015, the UK government proposed that all coal-fired power stations be phased out within ten years. Coal is a fossil fuel. Discuss the advantages and disadvantages of fossil fuels as energy resources. You may wish to refer to the following in your answer:

renewable   **energy per kilogram**   **cost**

carbon dioxide emissions (*6 marks*)

8 An engineer is designing a machine to lift patients in a hospital.

a The machine must increase the gravitational potential energy store for the patient by at least 1200 J in 20 s.

Calculate the minimum power needed from the motor and include the unit for power. The formula for calculating power is:

$$power = \frac{energy}{time}$$ (*2 marks*)

b The engineer needs to select an appropriate power source. The motor operates at 24 V.

Calculate the current needed and include the unit of current. The formula that relates current to electrical power is:

$$current = \frac{power}{potential\ difference}$$ (*2 marks*)

c For convenience, the engineer decides to use a rechargeable battery. The hospital managers need to know how much it will cost to charge the battery. The battery recharges at 0.060 kW for 2 hours, and the electricity company charges 15p per kW h.

Calculate the cost. The formula for cost is:

cost (p) = power (kW) × time (h) × cost per unit (p / kW h) (*2 marks*)

9 **Figure 5** shows the same gas at different temperatures.

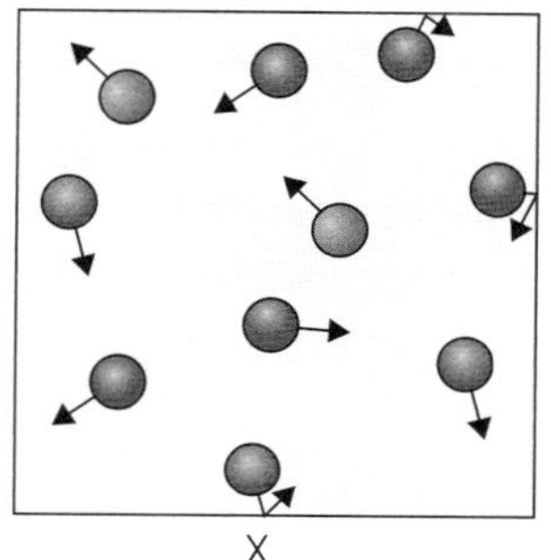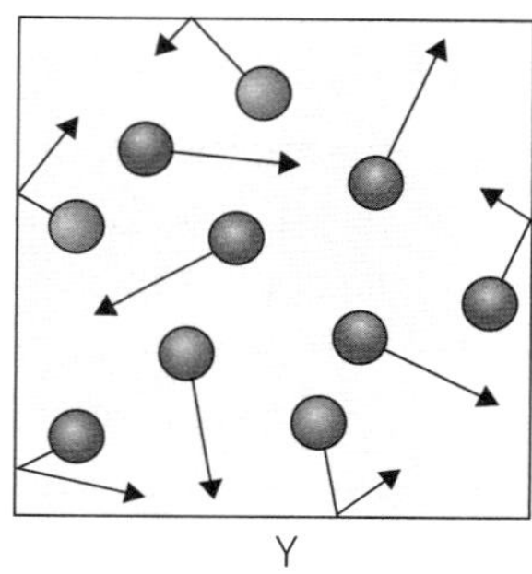

Figure 5

a i Which of these images shows a hot gas and which shows a cold gas?

Hot gas _______________

Cold gas _______________ (*1 mark*)

ii Decide which of these images shows the gas at a higher pressure. _________ (*1 mark*)

b i In a room at room temperature, 23 °C, two balloons are refilled with air. One is filled with air at 23 °C and the other air at 80 °C.

Explain why the balloon filled with hot air will rise, whereas the balloon filled with colder air will not. (*2 marks*)

ii Draw a forces diagram for a balloon that is rising. (*1 mark*)

10 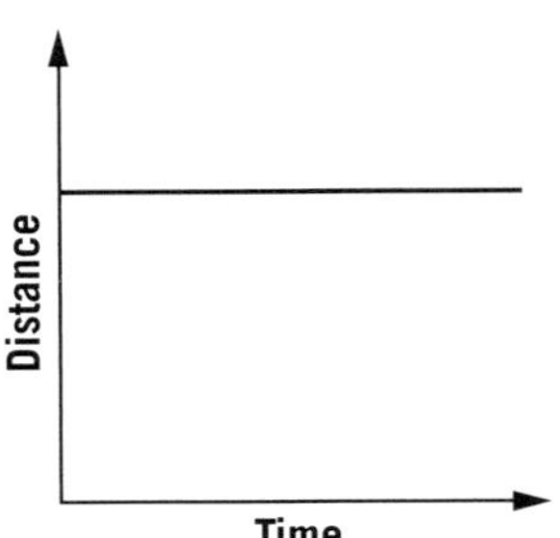 The pressure of a gas is important; for instance, air pressure helps us to predict the weather.

a Tick the variables that affect the pressure of a gas. *(2 marks)*

1 temperature ☐

2 electrical charge ☐

3 volume ☐

4 distance ☐

b Give the formula for calculating pressure. *(1 mark)*

____

c Sometimes people get stuck in thick mud by rivers. A problem for firefighters when they rescue them is avoiding get stuck themselves. A firefighter has a large board to stand on, which measures 1.8 m long and 0.50 m wide.

i Calculate the area of the board. *(1 mark)*

____

ii The firefighter has a weight of 800 N. Calculate the pressure he will exert when standing on the board and include the unit. *(2 marks)*

____

iii If the pressure on the ground is above 20 000 N/m², the firefighter will sink into the mud. Explain whether he will sink if he stands on the board. *(1 mark)*

____

11 **a** **Figure 6** shows a distance–time graph for one part of a journey.

Figure 6

Describe what is happening. *(1 mark)*

____

b **Table 2** contains data for Clara walking to visit a friend.

Table 2

Time (s)	Distance (m)
0	0
120	180
180	180
300	280

Draw a distance–time graph for this journey. *(3 marks)*

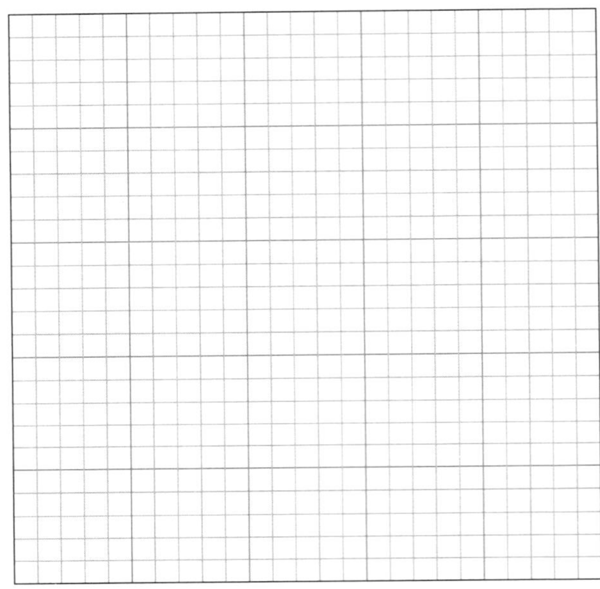

c Calculate the speed of the first part of Clara's journey. *(3 marks)*

____

12 Male stag beetles have jaws 30 mm long and a strong bite of 8.0 N. A biologist studying the beetle wants to understand how they bite so strongly. Work in metres.

Calculate the moment of force and include the units. The formula for moment of force is:

moment = force × distance from pivot *(2 marks)*

____

P2 Checklist

Revision question number	Outcome	Topic reference	☹	😐	☺
1a	State some uses of electromagnets.	P2 1.8			
1b	Describe some uses of electromagnets.	P2 1.8			
2a	State some properties of infrared radiation.	P2 2.5			
2b	State how energy and temperature are measured.	P2 2.3			
3a	Name the equipment used to measure potential difference.	P2 1.3			
3b	Describe how to measure potential difference.	P2 1.3			
4a	Name what flows in a circuit.	P2 1.2			
4b	Calculate resistance of a component and of a circuit.	P2 1.5			
4c	Describe the difference between conductors and insulators in terms of resistance.	P2 1.5			
5a	Identify the pattern of current in series and parallel circuits.	P2 1.4			
5b	Describe how current and potential difference vary in series and parallel circuits.	P2 1.4			
6a	Calculate work done.	P2 2.8			
6b	Interpret data on food intake for some activities.	P2 2.1			
7	State one advantage and one disadvantage of fossil fuels.	P2 2.6			
8a, b, c	Predict the power requirements of different equipment and how much it costs to use.	P2 2.7			
9a	Describe what happens when you heat up solids, liquids, and gases.	P2 2.3			
9b	Explain why some things float and some things sink, using force diagrams.	P2 3.4			
10a	State two things that can affect gas pressure.	P2 3.3			
10b	State the equation of pressure.	P2 3.5			
10c, d, e	Calculate pressure.	P2 3.5			
11a	Use a distance–time graph to describe a journey qualitatively.	P2 3.2			
11b	Plot data on a distance–time graph accurately.	P2 3.2			
11c	Calculate speed from a distance–time graph.	P2 3.2			
11d	Interpret distance–time graphs.	P2 3.2			
12	Calculate the moment of a force.	P2 3.6			

Answers

B1.1

A lipid, protein, vitamin, carbohydrate

B protein – fish; vitamins and minerals – fruit; carbohydrate – pasta; lipid – butter

C carbohydrate – main source of energy; lipid – store of energy, keep you warm, protect organs; protein – growth and repair of body tissues; water – needed in all cells and body fluids; vitamin – needed in tiny amounts to keep you healthy

D 1 false; 2 true; 3 true; 4 false

What you need to remember

balanced, nutrients, carbohydrates, lipids (either order), proteins, vitamins, minerals (either order), fibre

B1.2

A the solution changes colour

B pestle and mortar; pipette; filter paper; water bath

C filter paper, to the light, goes translucent

D starch – orange iodine solution – blue-black colour; sugar – blue Benedict's solution – brick-red colour; protein – blue copper sulfate and sodium hydroxide solution – purple colour; lipid – ethanol – cloudy, white layer

What you need to remember

food tests, iodine, starch, red, purple, protein, cloudy

B1.3

A energy, joules, different, more

B 3, 2, 4, 1

C obesity – eating too much food or too many fatty foods; deficiency – lack of vitamin or mineral; starvation – eating too little food

D obese – diabetes, heart disease; underweight – lack of energy, poor immune system

What you need to remember

malnourishment, underweight, deficiency, starvation, fat, obese

B1.4

A large, broken down, small

B clockwise from top: gullet, stomach, large intestine, small intestine, liver

C small intestine, anus, stomach, gullet, large intestine

What you need to remember

digestive system, gullet, stomach, acid, digestion, small, large, rectum, anus

B1.5

A protease, carbohydrase, lipase

B large intestine

C 1 false; 2 true; 3 true; 4 false

D carbohydrase – carbohydrate – sugar; protease – protein – amino acid; lipase – lipid – fatty acids and glycerol

What you need to remember

bacteria, vitamins, enzymes, catalyst, carbohydrase, sugar, protease, protein, lipase, glycerol, bile

B1.6

A medicinal – antibiotic, paracetamol, aspirin, ibuprofen; recreational – ecstasy, caffeine, alcohol, tobacco

B 1 medicinal; 2 recreational; 3 recreational; 4 medicinal

C alcohol – slows down the nervous system and damages the liver; tobacco – significantly increases risk of lung cancer and heart disease; caffeine – speeds up the nervous system

D headaches, anxiety, excess sweating

What you need to remember

drugs, recreational, medicinal, addiction, withdrawal symptoms

B1.7

A feeling relaxed and happy, difficulty walking and talking (drunk), unconsciousness, death

B liver, brain

C ethanol, bloodstream, nervous, depressant, slows down

D 1 true; 2 true; 3 false; 4 true

What you need to remember

ethanol, nervous, depressant, liver, unit, alcoholic(s)

B1.8

A lung cancer, heart attack, stroke

B nicotine, tar, nicotine, tar, carbon monoxide

C low birth weight baby, miscarriage during pregnancy

What you need to remember

cancer / disease, heart, passive, airways, monoxide, oxygen, stimulant, miscarriage

B2 Chapter 1 Pinchpoint

A this is an incorrect answer – enzymes are **not** living and are **not** used up during a reaction

B this is an incorrect answer – carbohydrates are **not** broken down into amino acids, they are broken down into sugar molecules

C this is an incorrect answer – the **only** enzymes found in saliva are carbohydrases, which break down carbohydrates

D this is the correct answer

Pinchpoint follow-up

A Enzymes are catalysts. This means they are not used up in a reaction; Enzymes are catalysts. This means they speed up a reaction; Enzymes are made of protein molecules; Enzymes are not living as they cannot respire.

B carbohydrate – carbohydrase → individual squares drawn and labelled as sugar molecules; protein – protease → individual hexagons drawn and labelled as amino acids; lipid – lipase → rectangles drawn (some shaded) and labelled as fatty acids and glycerol molecule

C carbohydrase – mouth, stomach, small intestine; protease – stomach, small intestine; lipase – small intestine

"

D many stains on clothes are food stains, enzymes will break down large food molecules (insoluble) into smaller molecules (soluble), easier to wash away; named example stated, e.g. grease stain is caused by lipids – lipase will break down lipid into fatty acids and glycerol

B2.1

A chloroplast
B glucose, oxygen
C water, light, oxygen
D carbon dioxide – enters through tiny holes on the underside of the leaf; water – diffuses into root hair cells; light – absorbed by chlorophyll in chloroplasts

What you need to remember

algae, producers, photosynthesis, consumers, water, glucose, chlorophyll

B2.2

A (clockwise from top-left) waxy layer, chloroplast, palisade layer, spongy layer, air space, guard cell, stoma
B chloroplast – contains chlorophyll to trap sunlight; stomata – allow gases to diffuse into and out of leaf; guard cells – open and close stomata; veins – transport water to cells in leaf; waxy layer – reduces water loss through evaporation
C 1 false; 2 true; 3 false; 4 true

What you need to remember

leaves, stomata, carbon dioxide, oxygen, xylem, palisade, chloroplasts

B2.3

A nitrates, potassium
B NPK fertilisers
C nitrate – healthy growth – poor growth, older leaves are yellowed; phosphate – healthy roots – poor root growth, younger leaves look purple; potassium – healthy leaves and flowers – yellow leaves with dead patches; magnesium – making chlorophyll – plant leaves turn yellow
D soil, soil, root hair cells, xylem, amino acids, proteins

What you need to remember

minerals, magnesium, potassium, nitrates, phosphates, deficiency, fertilisers

B2.4

A bacteria   **B** mutualistic
C a chemical; glucose; carbon dioxide; dark
D sulfur bacteria – near volcanic vents at the bottom of the sea – hydrogen sulfide; nitrogen bacteria – plant roots – nitrogen compounds

What you need to remember

bacteria, chemosynthesis, chemical, glucose, sulfur, sea

B2.5

A respiration
B glucose, oxygen

C oxygen, carbon dioxide
D labelled arrow added pointing to a mitochondrion

What you need to remember

aerobic respiration, glucose, mitochondria, water, carbohydrates, plasma, haemoglobin

B2.6

A lactic acid
B glucose → lactic acid
C aerobic: glucose is a reactant; oxygen is a reactant; carbon dioxide is produced; water is produced; transfers more energy per glucose molecule
anaerobic: glucose is a reactant; lactic acid is produced
D 1 true; 2 false; 3 false; 4 true

What you need to remember

anaerobic respiration, lactic acid, oxygen debt, fermentation, ethanol

B2.7

A food chain – a diagram that shows the transfer or energy between organisms; food web – a diagram showing linked food chains; producer – an organism that makes its own food; consumer – an organism that eats other organisms to gain energy
B **a** acacia tree / grass
   b impala / giraffe / zebra
   c cheetah / leopard / lion
   d cheetah / leopard / lion
   e impala / giraffe / zebra / cheetah / leopard
C food chain beginning with a plant with 3 correct links: grass / acacia → impala → cheetah / leopard → lion

What you need to remember

food chain, energy, producer, photosynthesis, consumers, prey, predator, food web, linked / interconnected

B2.8

A depend, interdependence, shelter, species, population, increase
B bioaccumulation
C **a** decrease – die from the disease
   b increase – fewer ladybirds eating them

What you need to remember

interdependence, population, increase, bioaccumulation

B2.9

A ecosystem – the living organisms in a particular area and the habitat in which they live; niche – particular place or role that an organism has in an ecosystem; co-exist – different types of organism living in the same place at the same time; habitat – the area where an organism lives; community – the organisms in an ecosystem

B quadrat

C **a** food sources / niches; parts of the tree

b habitat: oak tree; community: birds, ants, squirrels, woodlice, slugs, and oak tree

What you need to remember

habitat, community, ecosystem, co-exist, niche

B2 Chapter 2 Pinchpoint

A this is an incorrect answer – this is the word equation for **photosynthesis**

B this is an incorrect answer – aerobic respiration **also** occurs in plants and microorganisms

C this is the correct answer

D this is an incorrect answer – this is what happens when you breathe; respiration is **a chemical reaction**

Pinchpoint follow-up

A **a** glucose and oxygen underlined

b carbon dioxide and water circled

c glucose + oxygen (either order) → carbon dioxide + water (either order)

B **a** animal – aerobic respiration: plant – aerobic respiration, photosynthesis; microorganism – aerobic respiration, some species can photosynthesise

b respire, movement, growth, plants, bacteria, photosynthesise, glucose

C **a** to contract to cause movement

b to move tail so it can 'swim' to egg

c to enable it to move by spinning flagella; move towards light (detected by eye spot) to maximise photosynthesis

D glucose, digestion, small, diffuses, inhale, breathing, alveoli, carbon dioxide, exhaled

B3.1

A adaptations

B food, space to hunt, water, mates

C light – for photosynthesis; water – for photosynthesis and to keep their cells rigid; space – so roots can absorb enough water and leaves can absorb enough light; minerals – to make chemicals needed for healthy growth

D **a** to stop it being eaten **b** to spot prey / predator

c to get away from predator / catch prey

d to keep warm **e** to survive drought

What you need to remember

resources, mates, food, light, minerals, adaptations

B3.2

A warm, hibernation, birds, migration, thicker fur

B X, Y, Z

C **a** solid line – rabbits; dashed line – foxes

b Y, Z

What you need to remember

environment, leaves, fur / wool / coats, adapted, interdependence, increases, food

B3.3

A species, characteristics, offspring, identical, variation

B **two** from: size, length, width, number of spikes, stalk (petiole) length

C inherited – eye colour, blood group; environment – accent, pierced ears; combination of both – height, skin colour

What you need to remember

species, variation, inherited, environmental

B3.4

A discontinuous variation; continuous variation

B **a** continuous variation

b discontinuous variation

C **a** histogram

b bar chart

D continuous – leg length, fish mass, leaf surface area; discontinuous – blood group, flower colour, number of spots

What you need to remember

discontinuous, bar chart, continuous, histogram

B3.5

A DNA; chromosomes; genes

B gene – V; chromosome – U; cell – S; nucleus – T

C chromosomes, half, half, an egg, a sperm, fertilisation, 46

What you need to remember

nucleus, DNA, chromosomes, genes

B3.6

A changes in species over time

B pale, camouflaged, dark, pale, increasing, dark, pale, decreasing, increasing, dark

C 4, 6, 2, 5, 1, 3

D fossils, DNA

What you need to remember

evolved, millions, natural selection, survive, genes, fossils

B3.7

A biodiversity, endangered, extinct

B gene banks, captive breeding

C outbreak of disease – whole population killed by a microorganism; prolonged drought – lack of water leads to death of whole population; introduction of new predators – whole population eaten before they have the chance to reproduce successfully; introduction of new competitors – lack of food / water causing death of whole population; deforestation – loss of food source or shelter leads to death of whole population

D genetic samples, low, individuals

What you need to remember

biodiversity, habitats, disease, extinct, endangered, seed bank

B2 Chapter 3 Pinchpoint

A this is an incorrect answer – the caribou population is **always** higher than the wolf population

B this is an incorrect answer – when the predator population increases the prey population **decreases**

C this is an incorrect answer – when the prey population decreases the predator population **decreases**

D this is the correct answer

Pinchpoint follow-up

A **a** 2008   **b** 5300 wolves   **c** 280 000 caribou

B **a** increases, decreases, decreases, increases
   b **i** predator population increases a period of time later
     ii predator population decreases a period of time later

C 4, 1, 5, 3, 2

D the population of wolves could decrease, as they would be less successful at catching their food; as there are now fewer predators, the caribou population will increase to higher levels than previously seen; if wolves don't adapt to change, eventually none may be present in that area; if wolves also evolve to be able to run faster, the cycle will continue in a similar way to the existing trend

B2 Revision questions

1 **a** tomato plant, green fly, sparrow, hawk [3 for all correct, 2 for 2 correct, 1 for 1 correct]
   b energy [1]

2 **a** caffeine / tobacco / ethanol (alcohol) [1]
   b paracetamol / antibiotic [1]
   c tobacco [1]
   d ethanol (alcohol) [1]

3 DNA, nucleus, chromosomes, genes, characteristic [4 for all correct, 3 for 3 correct, 2 for 2 correct, 1 for 1 correct]

4 **a** adaptations [1]
   b to keep warm [1]; for camouflage / to stay hidden [1]

5 **a** stomach – B [1]; gullet – A [1]; large intestine – D [1]
   b large molecules are broken down [1] into smaller molecules [1]

6 **a** **two** from: carbohydrates [1], lipids [1], proteins [1], vitamins [1], minerals [1], fibre [1]
   b e.g. heart disease / stroke / diabetes / some cancers [1]
   c poor immune system / tired / lack of energy [1]

7 **a** glucose [1], oxygen [1]
   b chloroplast [1]
   c **i** break down cell walls [1] so chlorophyll escapes / to decolourise the leaf [1]
     ii iodine turns from orange-yellow [1] to blue-black if starch is present [1]
   d ethanol is flammable / can catch fire [1]

8 **a** **i** reduce water evaporation [1]
     ii stops cactus being eaten [1]
     iii collect water from a large area [1]
   b **i** loses leaves / enters dormant phase [1]
     ii hibernation / migration / grow thick fur [1]

9 **a** increase [1]   **b** decrease [1]

c toxic chemical taken into thrush when eat snails [1]; chemical taken into body of hawk when thrushes eaten [1]; as hawks eat many thrushes, level becomes so high it kills / bioaccumulation [1]

10 **a** protein – to repair body tissues and make new cells; carbohydrates – main source of energy; lipids – to provide a store of energy, insulation and protect organs [2 for all correct, 1 for 1 correct]
   b adds bulk to food [1] to keep it moving through intestine / help waste be pushed out of body / prevent constipation [1]
   c $\dfrac{2520}{8400} \times 100$ [1] = 30% [1]
   d add Benedict's solution [1], heat (in water bath) [1]; if solution turns orange-red it contains sugar [1]

11 **a** **i** carbohydrase / amylase [1]
     ii breaks down carbohydrates / starch [1] into glucose / sugar molecules [1]
   b speed up reactions [1] without being used up [1]
   c live on fibre in the intestines [1]; make vitamins / vitamin K [1]

12 **a** **i** oxygen [1], water [1]   **ii** mitochondria [1]
   b **two** from: anaerobic doesn't need oxygen, aerobic does [1]; anaerobic produces lactic acid, aerobic does not [1]; aerobic transfers more energy per glucose molecule [1]; aerobic produces carbon dioxide and water, anaerobic does not [1]
   c **i** fermentation (anaerobic respiration in yeast) [1]
     ii microorganism / yeast [1]

13 **a** **i** e.g. hair colour, ear piercing [1]
     ii e.g. eye colour, blood group, attached / free-hanging earlobes [1]
   b **i** number of spikes labelled on *x*-axis and frequency on *y*-axis [1]; appropriate scale used [1]; 3 data points plotted correctly [1]; all data points plotted correctly [1]
     ii 10 [1]
   c **i** discontinuous [1]
     ii more light allows more photosynthesis [1]; leaves grow bigger [1] so could have more spikes [1]

14 **a** **one** resource and **one** reason from: water [1] – for photosynthesis / to stay upright [1]; space [1] – to collect light / water / minerals [1]; minerals [1] – for healthy growth [1]
   b animals cannot photosynthesise to make food [1], and have to gain energy by eating other organisms [1]
   c **six** from: organisms in prey species show variation [1]; those most adapted survive **and** reproduce [1]; named adaptation, e.g. fastest [1]; less well-adapted die [1]; genes from most-adapted individuals are passed onto next generation [1]; offspring are likely to display the advantageous characteristic / advantageous characteristic becomes more common [1]; reference to natural selection [1]; process repeated over many generations [1]; over time can lead to the development of a new species [1]

C1.1

A from left – metals, non-metals

B properties of a typical non-metal – gas at room temperature, poor conductor of heat, brittle

C W, Y, Z

D non-metals, left, right, high, good, sonorous, physical, chemical

What you need to remember

non-metals, left, right, high, good, sonorous, physical, chemical

C1.2

A bars should be correctly and neatly plotted

B top, bottom, decreases, below, 29

C from top to bottom of Group 3, density increases; from top to bottom of Group 4, density increases; the pattern in density is similar for Group 3 and Group 4; from bottom to top of Group 4, density decreases

What you need to remember

groups, periods, groups, periods, group

C1.3

A left – bubbles of gas; right, from top – purple flame, potassium, water and universal indicator

B **a** one of the products of the reaction is **hydrogen** gas

   b the other product of the reaction is potassium **hydroxide**

   c potassium + water → potassium hydroxide + **hydrogen**

   d the universal indicator changes colour from green to purple, so the pH at the end could be **12 / 13 / 14**

   e rubidium is below potassium in Group 1, so the reaction of rubidium with water is **more** vigorous

C increases, less than, 0.26

What you need to remember

left, conduct, low, reactive, two / new, hydrogen, hydroxide

C1.4

A bottom symbol – corrosive – burns eyes – wear safety goggles
top symbol – toxic – difficulty breathing – use in a fume cupboard

B darker, top, iodine, solid, dark

C 1 true; 2 true; 3 false; 4 true

What you need to remember

right, halogens, metals, reactions, chloride, displaces

C1.5

A A, B, F

B 2, 4, 6

C Xe, Kr, Ar, Ne, He

D **a** Xe, Kr, Ar, Ne, He

   b balloons filled with argon, xenon, and krypton sink

What you need to remember

right, noble, metals, unreactive (inert)

C2 Chapter 1 Pinchpoint

A this is an incorrect answer – sodium is not the most reactive element in Group 1 and chlorine is not the most reactive element in Group 7

B this is an incorrect answer – potassium is less reactive than the elements below it in its group and chlorine less reactive than fluorine in its group

C this is the correct answer

D this is an incorrect answer – the patterns in reactivity are different for different groups of the Periodic Table: In Groups 1 and 2 the elements get more reactive from top to bottom; in Group 7 the elements get less reactive from top to bottom

Pinchpoint follow-up

A bottom, Group 1, most, top, Group 1, least, top, Group 7, most, bottom, Group 7, less

B 2, 4

C 2, 3, 4, 1

D lithium is at the top of Group 1 so it is the least reactive element in its group; fluorine is at the top of Group 7 so it is the most reactive element in its group; caesium is at the bottom of Group 1 so it is the most reactive element in its group; iodine is near the bottom of Group 7 so it is the least reactive element in its group

C2.1

A 1, 2, 3

B sand and water – with filter paper in a filter funnel; sand and steel nails – with a magnet; flour and marbles – with a sieve

C X and Z

D the same, one element, two, mixture, elements, compound, a compound

What you need to remember

pure, sharp, impure, temperatures, more, not, mixture, properties / melting points

C2.2

A A mixture of salt and water, when all the salt is dissolved, is called a **solution**.

In Brett's experiment, salt is the **solute or** In Brett's experiment, **water** is the solvent.

In Brett's experiment, water is the **solvent or** In Brett's experiment, **salt** is the solute.

B from left: sugar particle, water particle

C close together; in a regular pattern, separate from each other; surround, randomly, move around

D **W** – 109; **X** – 93; **Y** – 10; **Z** – 5

What you need to remember

solution, solute, solvent, solute, move, cannot, same, liquid, gas

C2.3

A solute – the solid or gas that dissolves in a liquid; solvent – the liquid in which a solid or gas can dissolve; soluble – if a substance dissolves in a solvent, the substance is __________ in that solvent; saturated solution – a solution in which no more solute can dissolve; solubility – the mass of solute that dissolves in 100 g of water to make a saturated solution

B **a** smooth curve drawn through, or close to, all the points
   b 33, 60, 240, increases

C 3, 4, 2, 5, 1

What you need to remember

saturated, solubility, soluble, soluble

C2.4

A (clockwise from left) residue, filter paper cone, filter funnel, clamp, conical flask, filtrate

B sand and water: residue – sand, filtrate – water; pieces of dirt from oil in a car engine: residue – pieces of dirt, filtrate – oil; coffee solution from ground-up coffee beans: residue – ground-up coffee beans; filtrate – coffee solution

C 1 – put a filter funnel in a **conical flask**; 2 – **do not** make a hole in a centre of the filter paper and fold it to make a cone; 3 – put the filter paper cone in the **filter funnel**; 4 – pour the mixture **through the filter paper**

D filter paper – has tiny holes in it; water particles are smaller than the tiny holes, so they pass through the filter paper; grains of sand are bigger than the tiny holes, so they stay in the filter paper cone

What you need to remember

filtering, insoluble, dissolved, liquid, filtrate, residue

C2.5

A from top – evaporation, distillation, evaporation, distillation

B 3, 5, 1, 4, 2

C (clockwise from top-left) thermometer, tap water out, condenser, beaker, pure water, tap water in, Bunsen burner, inky water

D heat, solvent, gas, condenses, liquid, beaker

What you need to remember

evaporation, solid, solute, distillation, condensation, solution

C2.6

A 2, 3, 4

B red: this colour ink is more soluble in water **and** this colour ink is attracted less strongly to the paper; blue: this colour ink is less soluble in water **and** this colour ink sticks more strongly to the paper

C two, carotene, green, xanthophyll

What you need to remember

solvent, chromatography, different, spot

C2 Chapter 2 Pinchpoint

A this is an incorrect answer – the mass of any solution is equal to the mass of solvent (water) **plus** the mass of solute (sugar or coffee)

B this is an incorrect answer – the mass of both solutions is equal to the mass of solvent (water) **plus** the mass of solute, **whether** the solute is coffee or sugar

C this is an incorrect answer – the mass of both solutions is equal to the mass of solvent (water) **plus** the mass of solute (sugar or coffee)

D this is the correct answer

Pinchpoint follow-up

A 1, 3, 5

B brown, still, does not change, does not change, plus, colourless, still, does not change, does not change, plus

C change, bean, equal, total, same, equal, sugar, water (either order)

D W – 106; X – 10; Y – 96; Z – 8

C3.1

A magnesium, zinc

B lighted, test tube, test tube, squeaky, oxygen, water

C **a** magnesium – many bubbles quickly formed; zinc – bubbles formed, but fewer than for magnesium; iron – bubbles formed, but slightly fewer than for zinc

   b All the metals reacted with dilute acid, ~~except for one~~. The reaction with ~~zinc~~ **magnesium** was most vigorous. The reaction with ~~iron~~ **lead** was least vigorous. When we tested the gas in the bubbles, we found out that it was ~~oxygen~~ **hydrogen** gas.

D **a** hydrogen   **b** chloride
   c iron chloride   **d** lead chloride + hydrogen

What you need to remember

metals, electricity, hydrogen, magnesium

C3.2

A iron oxide; lead oxide; magnesium oxide; zinc oxide

B magnesium, most, magnesium oxide, do not, oxide, unreactive, less

C zinc, iron, lead

D **a** s, g, s   **b** s, g, s   **c** s, g, s

What you need to remember

air, reactive, oxide, oxide, unreactive, state, solid, liquid

C3.3

A clockwise from left – bubbles of hydrogen gas; sodium; water with sodium hydroxide dissolved in it

B reactants – sodium and water; products – sodium hydroxide and hydrogen

C (most) potassium, sodium, lithium (least)

D caesium – explosive reaction; calcium – vigorous reaction, with rapid bubbling; magnesium – tiny bubbles form very slowly; gold – no reaction

What you need to remember

metals, top, unreactive, hydrogen, hydroxides

C3.4

A **a** magnesium   **b** aluminium
   c zinc   **d** iron

B more, displace, copper, sulfate

C 1, 2, 5

D **a** magnesium sulfate   **b** lead

   c copper   **d** magnesium (oxide), iron

What you need to remember

more, less, iron oxide, oxide, iron

C3.5

A cannot, the same, more, strongly

B **a** carbon

   b copper, iron, lead, zinc

   c aluminium is above carbon in the reactivity series

C 2, 1, 3

D **a** $100 - 20 = 80\%$

   b $\dfrac{20}{100} \times 100\,\text{kg} = 20\,\text{kg}$

   c $\dfrac{20}{100} \times 2000\,\text{kg} = 400\,\text{kg}$

What you need to remember

crust, mixed, ore, displacement / chemical, displaces, iron

C3.6

A 1, 2, 4

B making walls – do not react with water, strong when forces press on it; making jet engine blades – do not react with water, high melting point; making plates and cups – do not react with water, high melting point; insulators for power lines – do not conduct electricity, do not react with water

C the mass that makes the ceramic break – dependent variable; the position of the masses – control variable; the ceramic material – independent variable; the size / thickness of the piece of ceramic material – control variable

What you need to remember

metal, metal, high, insulators, physical, alkalis

C3.7

A **a** a polymer is a substance with very **long** molecules

   b a polymer molecule has many **identical** groups of atoms, joined together in a long chain

   c there are **many** different polymers

   d all polymers have **different** properties

B poly(propene) is used to make ropes because it is strong and flexible; nylon can be used to make artificial heart valves because it is strong and flexible, does not wear away, and is not damaged by blood; cotton is used to make clothes because it can be spun into flexible threads that can be woven into cloth; Kevlar® is used to make bullet-proof vests that police officers can wear all day because it is strong and lightweight

C PETE and rigid PVC because both are waterproof and rigid

What you need to remember

long, many, many, different, atoms

C3.8

A 1, 3, 4, 5

B **a** aeroplane wing – light and strong; heat shields for spacecraft – not damaged at high temperatures; hockey stick – light and strong; brake discs in high-performance cars – not damaged at high temperatures

   b **i** light and strong   **ii** heat shields for spacecraft

     iii brake discs in high-performance cars

C **a** bars for wood, bamboo, and glass drawn accurately

   b e.g. poly(propene) reinforced by wood and bamboo is less strong than poly(propene) reinforced by silk; the strongest material is made from poly(propene) and glass

What you need to remember

mixture, different, combination, pushing / squashing, pushing / squashing

C2 Chapter 3 Pinchpoint

A this is an incorrect answer – the reactivity series shows that lead is **less** reactive than zinc, so lead cannot displace zinc from its compounds

B this is an incorrect answer – a less reactive metal cannot displace a more reactive metal from its compound. Since the metal in the compound (lead in lead nitrate) is the more reactive metal, there is no reaction

C this is the correct answer

D this is an incorrect answer – in a displacement reaction, a more reactive metal displaces a less reactive metal from its compound. The reactivity series shows that magnesium is **less** reactive than sodium, so the reaction cannot occur

Pinchpoint follow-up

A **a** lead   **b** sodium   **c** zinc

   d sodium   **e** magnesium   **f** calcium

   g silver

B **a** 1, 2, 4, 5

   b **Zinc** is more reactive than **lead**. **Magnesium** displaces **zinc** from its compounds. **Calcium** displaces **magnesium** from its compounds. **Silver** displaces **gold** from its compounds.

C react, more, displaces, more, does not displace, less, does not displace, more, displaces, more, does not displace, less

D **a** magnesium, lead, zinc, magnesium, sodium, magnesium

   b pairs that react: magnesium and copper oxide, zinc and lead oxide; the other pairs do not react

C4.1

A from top – crust, mantle, outer core, inner core

B core – rocky, of thickness 8 km to 40 km; mantle – mainly solid rock, but can flow; inner core – solid, mainly iron and nickel; outer core – liquid, mainly iron and nickel

C **S** – 1, 2, 5; **D** – 3, 4

What you need to remember

layers, inner, outer, mantle, solid, flow, crust, atmosphere, troposphere, nitrogen, oxygen

C4.2

A   tiny, porous, not very strong, soft

B   1, 4, 2, 3

C   physical weathering – the breaking up of rock because of changing temperature; erosion – the breaking of a rock into sediments, and their movement away from the original rock; chemical weathering – the breaking up of rock by the action of substances, for example those in rainwater; deposition – the settling of sediments that have moved away from their original rock; cementation – the 'gluing together' of sediments by different substances to make sedimentary rocks; compaction – squashing sediments together to make new rock by the weight of layers above; transport – the movement of sediments far away from their original rock

What you need to remember

metamorphic, soft, biological, sediments, transport, deposition, compaction, cementation (either order), compaction, joins, cementation

C4.3

A   basalt – igneous; granite – igneous; limestone – sedimentary; marble – metamorphic; sandstone – sedimentary; slate – metamorphic

B   **a** both   **b** both   **c** igneous   **d** metamorphic   **e** igneous

C   freezes, slowly, more, bigger, igneous, big, igneous, small

What you need to remember

magma, lava, solidifies / freezes, crystals, non-porous, hard, pressure, crystals

C4.4

A   solid rock of any type – melts – to make lava or magma; solid rock of any type – is broken into smaller pieces – to make sediments; solid rock of any type – has its particles rearranged – to make metamorphic rock; liquid rock – freezes – to make igneous rock; sediments of any rock type – join together – to make sedimentary rock

B

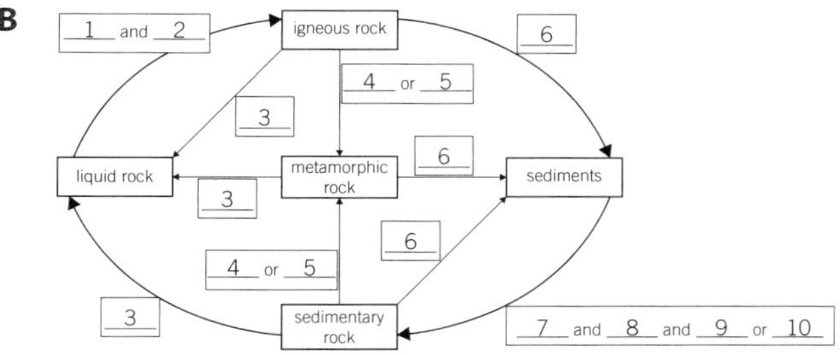

What you need to remember

recycled, sedimentary, freezes / solidifies, igneous, temperatures / heating, metamorphic, uplift

C4.5

A   oceans, fossil fuels, the atmosphere, some sedimentary rocks

B   atmosphere, did not change, increased, faster

C   **removes** CO_2 from the atmosphere – photosynthesis by plants, the formation of some sedimentary rocks, dissolving in the oceans, the formation of coal and other fossil fuels; **adds** CO_2 to the atmosphere – respiration by plants, respiration by animals, burning petrol in cars, burning diesel in cars

What you need to remember

reservoirs, sedimentary, fossil, cycle, respire, burn / combust, photosynthesis, dissolving, same, change / increase / decrease

C4.6

A   global warming; climate change; deforestation; greenhouse effect

B   deforestation – less carbon dioxide is removed from the atmosphere; burning fossil fuels – more carbon dioxide goes into the atmosphere; every year, more carbon dioxide is added to the atmosphere than is removed – the concentration of carbon dioxide in the atmosphere increases; climate change – rising sea levels, floods, and more hurricanes and droughts

C   1, 6, 3, 4, 2, 5

D   more, fossil, deforestation, reflected radiation, warmer / hotter, greenhouse, increasing, warming, climate, weather

What you need to remember

more, fossil, deforestation, radiation, warmer / hotter, greenhouse effect, higher / greater, global warming, climate, weather

C4.7

A   1, 2

B   1, 5, 4, 2, 3

C   advantages – 1, 3, 4; disadvantages – 2, 5

D   low-density poly(ethene) and poly(propene)

What you need to remember

processing, reduced / lower, pollution / noise / congestion

C2 Chapter 4 Pinchpoint

A   this is the correct answer

B   this is an incorrect answer – the particles in sedimentary rock are rearranged, but the answer does not describe what **causes** this rearrangement

C   this is an incorrect answer – metamorphic rock is formed by the action of **heat or pressure** on existing rock, not by the breaking down of existing rock into small pieces

D   this is an incorrect answer – melting is **not** involved in the formation of metamorphic rock

Pinchpoint follow-up

A   rock is broken up to make small pieces of rock – sedimentary; rock is heated. It does not melt, but its particles are rearranged – metamorphic; rock is heated. It melts and freezes again – igneous; rock experiences high pressures. Its particles are rearranged – metamorphic

B   2, 4, 1, 3 / 5 (either order)

C   **M** – 2, 4, 6; **S** – 1, 3, 5; **B** – 7

D   hot, does not melt, crystals, high, crystals

C2 Revision questions

1 nitrogen [1]
2 hydrogen [1] and sodium hydroxide [1]
3 salt from salt solution – evaporation [1]; water from salt solution – distillation [1]; undissolved salt from saturated salt solution – filtration [1]
4 low boiling point [1]
5 **a** bars correctly drawn [1]
  b from top to bottom of the group, melting point increases [1]
6 igneous – by freezing liquid rock [1]; metamorphic – by high pressures changing existing rock [1]; metamorphic – by high temperatures changing existing rock, without melting [1]
7 poly(propene) [1]
8 W, Y, and Z [1]
9 sugar – solute [1]; water – solvent [1]
10 X [1] and Z [1]
11 96 g – 85 g = 11 g [1]
12 apparatus – filter funnel with filter paper and conical flask / beaker [1]; fold the filter paper to make a cone, and place it in the filter funnel [1]; place the filter funnel above the conical flask / beaker [1]; pour the mixture into the filter paper cone [1]; the sand will remain in the filter paper cone [1]; the water will run down into the conical flask / beaker [1]
13 B [1]
14 **a** X [1]
   b type of metal [1]
   c temperature [1] and volume of acid [1]
15 mass $= \dfrac{35}{100} \times 200$ kg [1]
        $= 70$ kg [1]
16 3 [1] and 4 [1]
17 **a** the mass of solute that dissolves in 100 g of water to make a saturated solution [1]
   b as temperature increases, so does solubility [1]
   c **i** 95 g/100 g of water [1]   **ii** 114 g/100 g of water [1]
18 the rate at which animals and plants added carbon dioxide to the atmosphere by **respiration** [1] was balanced by the amount removed from the atmosphere by plants in **photosynthesis** [1] and **dissolving** in the oceans [1]

P1.1

A clockwise from top-left: proton, electron, neutron
B ⁺ve charges where they were; fewer ⁻ve charges on jumper; more ⁻ve charges on balloon
C **a** friction, electrons, electrons, positive, negative
  b A – repel; B – attract; C – attract; D – repel
  c the space around a charged object; which causes other charged objects to be attracted or repelled
D charged particles move from hair to balloon; both now have opposite charge to each other so they attract

What you need to remember

protons, electrons, neutral, friction, electrons, negatively, electric field, repel, attract, lightning

P1.2

A **a** movement of charge (or charged particles); amount of charge flowing per second
  b ammeter
B cell, pushing, complete, closing, switch, opening, switch
C **a**   **b** series

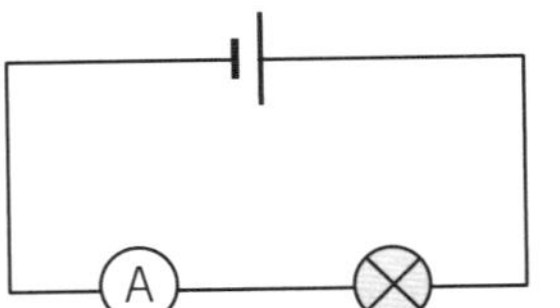

What you need to remember

electrons, second, ammeter, series, ammeter, amp, A, cell, battery (either order), motor, switch, complete

P1.3

A **a** volt, V   **b** cell, voltage, cell, volts
  c connect voltmeter into circuit so that it is connected across bulb in parallel
  d bulb and cell connected in one loop; voltmeter (V in circle) connected in extra parallel loop around bulb
B p.d. measured across the battery is 1.2 volts
C push from battery not enough to get current through phone / not enough energy transferred to phone

What you need to remember

potential difference, voltmeter, parallel, volts, V, voltage, rating, high / large, energy, low / small

P1.4

A In a series circuit, current is the same everywhere. In a series circuit, potential difference across each component adds up to that across the cell. In a parallel circuit, current through each branch adds up to that through the cell. In a parallel circuit, potential difference across each branch is the same as that across the cell.
B series – all components form one loop; parallel – more than one loop or branch
C **a** 6 V   **b** 0.8 A   **c** 6 V

What you need to remember

series, smaller, current, potential difference, potential difference, parallel, larger, current, current, potential difference

P1.5

A **a** a measure of the 'push' that is required to get current through a component (or potential difference across a component divided by the current through that component) **or** opposition to the flow of current
  b insulators: plastic, wood, glass; conductors: copper, gold
  c conductors have low resistance; insulators have high resistance
B $\dfrac{6.0}{0.20} = 30$ (Ω)
C **a** current (A) $= \dfrac{\text{potential difference (V)}}{\text{resistance } (\Omega)}$

b $\dfrac{9.0}{30} = 0.30$ A

D $0.50 \times 20 = 10$ V

What you need to remember

resistance, potential difference, current, ohm, resistance, high, low

P1.6

A a

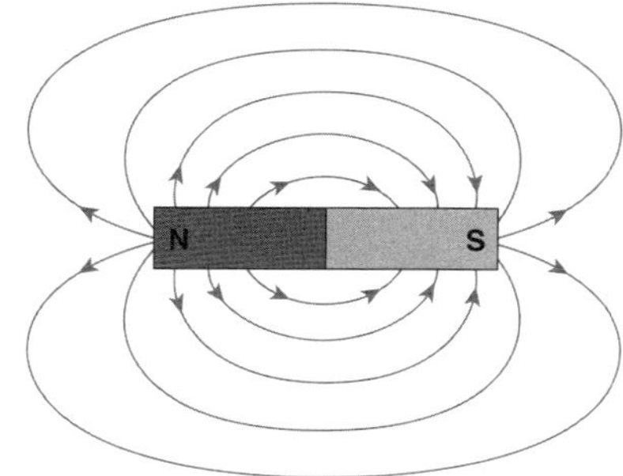

    b similar diagram to part **a**, but twice as many field lines

B a attract   **b** attract   **c** repel   **d** repel

C curved field lines like **A a**, pointing from magnetic south to magnetic north

What you need to remember

magnets, magnetic materials, iron, iron, north pole, south pole (either order), north, Earth's, south, different, same, magnetic field, magnetic field lines, more

P1.7

A a electromagnets can be turned off; permanent magnets cannot

    b wire in a coil surrounding a core, connected in series with a power supply (or labelled with current flowing through the coil) and a switch

    c wrap loops of wire; around a magnetic core such as iron; connect to a power supply so that current flows

B **two** from: increase number of turns on the coil; increase current flowing in the wire / increase the power supply potential difference; use a magnetic material in the core

C a independent: current through coil; dependent: number of paper clips picked up; control variables (**two** from): material in core, number of turns on coil, which type / size of paper clips used

    b e.g. build an electromagnet with an iron core to pick up steel paper clips; vary how much current flows through coil; count how many paper clips it picks up for each current

What you need to remember

coil, turns / loops, core, more turns, current, material, magnetise, permanent, turned off, field, bar

P1.8

A clockwise from top: magnets, coil, power supply

B 2, 1, 3, 4

C **two** from: to levitate a train; to propel train; to lift a car (or other heavy metal object); to sort metal for recycling; in a relay to control large current in an X-ray machine; in a relay to control large current to start a car

D from left: low current control, relay, high power device

What you need to remember

motor, coil, permanent magnets, current, electromagnet, spin / move, relay, larger, circuit, electromagnet, magnetised, completing / closing, safety

P2 Chapter 1 Pinchpoint

A this is the correct answer

B this is an incorrect answer – in a series circuit the current is the same everywhere

C this is an incorrect answer – a cell still has potential difference across it even if no current is flowing through it

D this is an incorrect answer – the current that a cell pushes around a circuit varies depending on the resistance of the circuit

Pinchpoint follow-up

A a current $= \dfrac{\text{potential difference}}{\text{resistance}}$ $\dfrac{3.0}{10} = 0.30$ A

    b current will decrease and bulb will be less bright

    c the bulb will probably blow/fail, so it will stop lighting and current will drop to zero

B a charge flowing per second

    b one unit   **c** 1 A   **d** 5 A

C a potential difference the battery provides – the amount of 'push' it gives to charges in the circuit

    b open, no current, a current, potential difference

    c **i** cell   **ii** battery of cells   **iii** cell

D a potential difference, connected, current, resistance

    b i current $= \dfrac{\text{potential difference}}{\text{resistance}} = \dfrac{3.0}{1.0} = 3.0$ A

      ii current $= \dfrac{\text{potential difference}}{\text{resistance}} = \dfrac{3.0}{20} = 0.15$ A

P2.1

A a apples and peas store similar amounts of energy; chips and chocolate store much more energy than apples and peas; chocolate stores the most; apples store the least

    b $20 \times 13 = 260$ kJ required for walk, chocolate supplies 1500 kJ, which is **more** than needed

B a sleeping 300; working 600; running 3600

    b sleeping involves no exercise but running involves a lot, so requires more energy

C swimmer eats nearly twice as much as the office worker; swimmer is doing a lot of exercise; office worker is doing a lot of sitting; swimmer uses far more energy than the office worker so needs to eat more to balance that

What you need to remember

energy, joule, kilojoule, activities, fat

P2.2

A **two** from: electric current, sound, light

B **a** chemical store – coal fuel (and oxygen in air); thermal store – water; thermal store – surroundings

   b chemical store – fuel in car (and oxygen in air); kinetic store – car; thermal store – car and surroundings

   c gravitational potential store – Charlie and scooter; kinetic store – Charlie and scooter; thermal store – scooter wheels and ground

C **a** chemical store of coal fuel (and oxygen in air) empties as coal used up; thermal store of water fills as it gets hotter; thermal store of surroundings fills as heat lost to surroundings

   b chemical store of fuel in car (and oxygen in air) empties as petrol used up; kinetic store of car fills as car moves and gets faster; thermal store of car and surroundings fills as both warm up

   c gravitational potential store of Charlie and scooter empties as Charlie goes downhill and has less height above the base of the hill; kinetic store of Charlie and scooter fills as speed increases; thermal store of scooter wheels and ground fills as friction makes wheels get hotter

What you need to remember

conservation, energy, energy stores, fill, total, chemical store, thermal, kinetic, gravitational potential, elastic, dissipated, thermal, thermal

P2.3

A °C, stays the same, increases, move / vibrate, J, thermal, increases

B **a** the water particles vibrate more, but do not move past each other

   b the water particles vibrate more and move past each other faster

C **a** soup – temperature decreases; air – temperature increases

   b hot object heats colder object until both reach same temperature

What you need to remember

temperature, thermometer, degrees Celsius, °C, same, equilibrium, no, more, thermal, mass, type (either order)

P2.4

A reduce, slower, insulators

B **a** particles heated, vibrate more; collide with neighbours and make them vibrate more; energy transferred from hotter place to colder place

   b particles heated, vibrate and move apart; liquid becomes hotter and less dense and rises; energy transferred from hotter place to colder place; colder more dense liquid sinks to take its place and be heated

C best conductor will transfer energy the fastest, so wax will melt fastest, recording shortest time; copper is the best conductor; glass is an insulator; aluminium is a conductor

What you need to remember

thermal, hot, cooler / colder / cold, vibrate, conduction, solid, gaseous / gas, dense, dense, convection, convection current

P2.5

A **a** black absorbs infrared better than white, so black clothes warm up and dry faster

   b matt surfaces absorb infrared better than shiny ones and black surfaces absorb more infrared than any other colour, so matt black surfaces heat the water faster

B **a** conduction – in solids, particles vibrating and colliding with neighbours; convection – particles moving from a hotter place to a colder place; radiation – emission and absorption of infrared radiation

   b hot objects emit infrared radiation; when an object absorbs this radiation, it causes heating

   c because radiation reaches the Earth from the Sun through space, where there are no particles / which is a vacuum

What you need to remember

conduction, convection, radiation, infrared radiation, thermal imaging camera, absorb, transmitted / absorbed / reflected (any order) × 3, black (or dark), matt (or dull), white (or light), shiny / gloss, black, matt

P2.6

A **a** renewable: wind, solar, geothermal, tidal, biomass, wave, hydroelectric
   non-renewable: coal, oil, natural gas, nuclear (uranium)

   b renewable continuously being remade; non-renewable cannot easily get more of them when they run out

B **a** natural gas, biogas

   b 2, 5, 1, 4, 6, 3

C renewable – will never run out; non-renewable – are often cheaper, more energy available for each amount of the resource

What you need to remember

energy resource, fossil fuels, non-renewable, thermal power station, renewable, carbon dioxide, sulfur dioxide

P2.7

A energy – J – cannot be created or destroyed
   power – W – how quickly stores are emptied and filled

B **a** $\dfrac{45000}{30} = 1500\,\text{W}$

   b **i** $10 \times 230 = 2300\,\text{W}$

   ii $2300 \times 30 = 69\,000\,\text{J or } 69\,\text{kJ}$

   c **i** $2 \times 4 = 8\,\text{kW h}$

   ii $2 \times 4 \times 15 = 120\text{p or £1.20}$

What you need to remember

energy, time, energy, power, current, potential difference (either order), power rating, watt, kilowatt, kilowatt hour

P2.8

A lever, smaller, greater, multiplier, smaller, conserves

B $750 \times 20 = 15\,000\,\text{J or } 15\,\text{kJ}$

C **a** $120 \times 0.40 = 48\,\text{J}$

   b 48 J (same as for **a**), because machines conserve energy

What you need to remember

work, energy, done, joule, force, work, energy, simple machines, levers, gears (either order)

P2 Chapter 2 Pinchpoint

A this is the correct answer; it is a correct statement of the law of conservation of energy

B this is an incorrect answer; the fuel undergoes a **chemical** change when it is burnt and its **chemical** store empties

C this is an incorrect answer; it is a wrong statement about the law of conservation of energy – the losses from emptying stores exactly match the gains from filling stores

D this is an incorrect answer; the soup undergoes a **thermal** change and its thermal store **fills**

Pinchpoint follow-up

A a dissipation, thermal store, minimum, lots, thermal store

   b i use a screen / put a lid on the saucepan

      ii less energy is dissipated to surroundings, so less fuel is used up to fill the thermal store of the soup

B a thermal – cooling soup – temperature decreases – empties; chemical – burning petrol in a car engine – mass of fuel reduced – empties; kinetic – accelerating bike – speed of bike increases – fills; gravitational potential – lifting bag of shopping – height of bag increases – fills; elastic – stretching elastic band – length of elastic band increases – fills

   b amount / volume / mass of fuel decreased

   c chemical

C a increases   **b** increases

   c the computer gets hot when in use   **d** thermal

   e appear / disappear, more / less, created, destroyed (either order), transferred

D a soup – temperature decreases; fuel – mass of fuel increases; lift – gravitational potential; car – speed increases – speed decreases

   b thermally

P3.1

A stationary, relative, compared, 25

B a $\dfrac{500}{30}$ = 1.7 m/s (to 2 sig. fig.)

   b $\dfrac{160}{2.0}$ = 80 km/h (to 2 sig. fig.)

   c $330 \times 12 = 4000$ m (to 2 sig. fig.)

C a the speed of one object compared to another

   b $25 - (-35) = 60$ m/s

What you need to remember

speed, distance, metre(s) per second, distance, speed, speed, instantaneous speed, average speed, relative motion, relative speed, relative speed

P3.2

A a stationary (not moving)

   b moving with constant speed

   c accelerating (speeding up)

B a time 0–10 s: horizontal line at 0 m; time 10–30 s: straight diagonal line up to 24 m; time 30–60 s: horizontal line at 24 m; time 60–80 s: straight diagonal line up to 72 m

   b car does not move for first 10 s; constant speed for 20 s; stops for 30 s; faster constant speed than before, for 20 s

   c speed = $\dfrac{\text{distance}}{\text{time}}$ = $\dfrac{24}{30-10}$ = 1.2 m/s

What you need to remember

distance–time, horizontal, speed, accelerating

P3.3

A a same number of particles packed into smaller space, with longer arrows

   b must withstand lots of collisions with air particles

B an increase in temperature – causes an increase in pressure – because the gas particles move faster, so collide harder and more often; an increase in volume – causes a decrease in pressure – because the particles have further to travel so collide less often

C a collisions between air particles, and between them and any surface in contact with the air

   b pressure decreases with height; less oxygen so harder to breath / can cause altitude sickness

What you need to remember

gas pressure, particles, compressed, density, atmospheric pressure, atmospheric pressure, smaller / less

P3.4

A a boat – float, pebble – sink, basketball – float

   b float: density of object < density of water; sink: density of object > density of water

B larger pressure the deeper you go; at surface, same as atmospheric pressure

C a upwards arrow from where ship touches water labelled upthrust (or buoyancy); downwards arrow from same place, same length, labelled weight (or gravity)

   b upwards short arrow from middle of anchor labelled upthrust (or buoyancy); downwards long arrow from same place labelled weight (or gravity)

What you need to remember

liquid pressure, pressure, upthrust, density, incompressible

P3.5

A a lying flat: $\dfrac{20}{0.12}$ = 170 N/m² (to 2 sig. fig.)

     on its end: $\dfrac{20}{0.060}$ = 330 N/m² (to 2 sig. fig.)

   b $\dfrac{3 \times 20}{0.12}$ = 500 N/m²

B **a** $\dfrac{5.0}{0.010} = 500$ N/m²

   b $\dfrac{2500}{6.4} = 390$ N/m² (to 2 sig. fig.)

C with narrow heel, person's weight is concentrated on small area so pressure is greater and marks made in wood; with wider heel, larger area for same weight so pressure is less and no marks in wood

What you need to remember

pressure, force, surface area, newtons per square metre, force, pressure, large, small, pressure, larger, less

P3.6

A **a** when an object is in equilibrium the sum of the clockwise moments is equal to the sum of the anticlockwise moments

B **a** $5.0 \times 0.50 = 2.5$ N m   **b** $50 \times 0.40 = 20$ N m

C centre of gravity is where their weight acts; if it off to the side they will fall that way

What you need to remember

pivot, moment, moment, force, distance, pivot, newton metre, force, pivot, moment, equilibrium, law of moments, weight, above, below (either order), no, topple

P2 Chapter 3 Pinchpoint

A this is an incorrect answer; Daisy is **moving** with a constant speed

B this is an incorrect answer; Daisy is accelerating and so her speed is **changing**

C this is the correct answer

D this is an incorrect answer; Daisy is moving with a constant speed, so is **not** accelerating

Pinchpoint follow-up

A **a** distance travelled, time taken, no / zero, zero, level

   b **i** horizontal line

      ii starting horizontal and curving upwards

B **a** slope, speed, steep, shallow, changing, curve

   b **i** straight, diagonal, shallow, upward line

      ii straight, diagonal, steep, upward line

      iii starting horizontal and curving upwards

C **a** 50 s to 100 s   **b** 100 s to 200 s   **c** 0 s to 50 s

D **a** unbalanced, change, constant, constant, cannot

   b

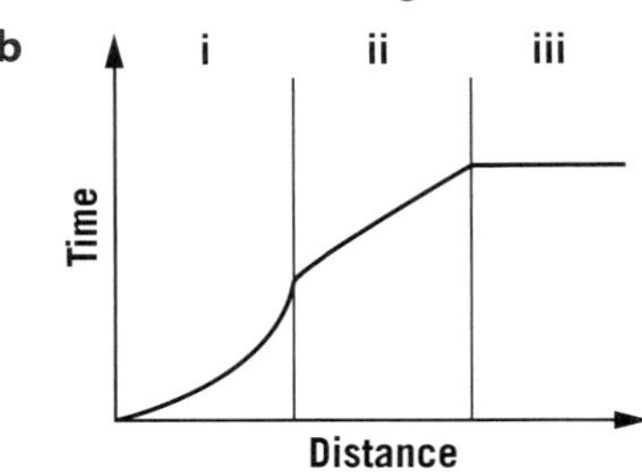

   c **ii** and **iii**

P2 Revision questions

1 **a** separating metal for recycling [1]; motor for electrically powered train [1]

   b separating metal – iron or steel attracted to electromagnet, other materials are not [1]; motor – magnetic field of electromagnet pushes against magnetic field of permanent magnet, gives force to move train [1] (**do not accept** electromagnet pushes against magnet)

2 **a** transmitted through the air; absorbed by the toast; emitted from the heating element

   b degree Celsius (°C); joule (J) / kilojoule (kJ) / kilowatt hour (kW h)

3 **a** circuit diagram for cell and bulb in series [1], with voltmeter in parallel with bulb [1]

   b voltmeter [1]

4 **a** charge [1]   **b** $\dfrac{12}{4.0} = 3.0$ [1] Ω [1]

   c wire – conductor [1]; wrapping – insulator [1]

5 **a** series: current same everywhere / all currents equal [1] parallel: sum of currents through branches / bulbs equal to current through supply / battery [1]

   b 15 V [1], 15 V [1]

6 **a** $200 \times 3000 = 600\,000$ [1] J [1]

   b larger [1], doing more work / exercise [1]

7 **six** from: disadvantages: non-renewable [1]; will run out eventually [1]; emit carbon dioxide when burnt [1]; produce pollutants such as sulfur dioxide, nitrogen oxides, and particulates [1]; cause climate change [1]
advantages: a lot of energy per kilogram [1]; low cost [1]

8 **a** $\dfrac{1200}{20} = 60$ [1] W [1]   **b** $\dfrac{60}{24} = 2.5$ [1] A [1]

   c $0.060 \times 2 \times 15 = 1.8$ [1] p [1]

9 **a** **i** hot: Y [1], cold: X [1]   **ii** Y [1]

   b **i** hot air inside the balloon is less dense than the air around / outside the balloon [1], so the air inside the balloon has a smaller mass and weight than the air around / outside the balloon [1]

      ii long upwards arrow from middle of balloon, short downwards arrow from same place [1]

10 **a** temperature [1], volume [1]

   b pressure $= \dfrac{\text{force}}{\text{area}}$ [1]

   c **i** $1.8 \times 0.5 = 0.90$ m² [1]

      ii $\dfrac{800}{0.90}$ [1] $= 890$ N/m² [1] (to 2 sig. fig.)

      iii no, he will not sink as $890 < 20\,000$ N/m² [1]

11 **a** standing still / stationary / not moving [1]

   b axes with reasonable scales [1], accurately plotted points [1], joined point-to-point with straight lines [1]

   c speed $= \dfrac{\text{distance}}{\text{time}} = \dfrac{180}{120}$ [1] $= 1.5$ [1] m/s [1]

12 $\dfrac{30}{1000} = 0.030$ m [1]; $8.0 \times 0.030 = 0.24$ N m [1]

Periodic table

1	2												3	4	5	6	7	0
							1 **H** hydrogen 1											4 **He** helium 2
7 **Li** lithium 3	9 **Be** beryllium 4												11 **B** boron 5	12 **C** carbon 6	14 **N** nitrogen 7	16 **O** oxygen 8	19 **F** fluorine 9	20 **Ne** neon 10
23 **Na** sodium 11	24 **Mg** magnesium 12												27 **Al** aluminium 13	28 **Si** silicon 14	31 **P** phosphorus 15	32 **S** sulfur 16	35.5 **Cl** chlorine 17	40 **Ar** argon 18
39 **K** potassium 19	40 **Ca** calcium 20	45 **Sc** scandium 21	48 **Ti** titanium 22	51 **V** vanadium 23	52 **Cr** chromium 24	55 **Mn** manganese 25	56 **Fe** iron 26	59 **Co** cobalt 27	59 **Ni** nickel 28	63.5 **Cu** copper 29	65 **Zn** zinc 30		70 **Ga** gallium 31	73 **Ge** germanium 32	75 **As** arsenic 33	79 **Se** selenium 34	80 **Br** bromine 35	84 **Kr** krypton 36
85 **Rb** rubidium 37	88 **Sr** strontium 38	89 **Y** yttrium 39	91 **Zr** zirconium 40	93 **Nb** niobium 41	96 **Mo** molybdenum 42	[98] **Tc** technetium 43	101 **Ru** ruthenium 44	103 **Rh** rhodium 45	106 **Pd** palladium 46	108 **Ag** silver 47	112 **Cd** cadmium 48		115 **In** indium 49	119 **Sn** tin 50	122 **Sb** antimony 51	128 **Te** tellurium 52	127 **I** iodine 53	131 **Xe** xenon 54
133 **Cs** caesium 55	137 **Ba** barium 56	139 **La*** lanthanum 57	178 **Hf** hafnium 72	181 **Ta** tantalum 73	184 **W** tungsten 74	186 **Re** rhenium 75	190 **Os** osmium 76	192 **Ir** iridium 77	195 **Pt** platinum 78	197 **Au** gold 79	201 **Hg** mercury 80		204 **Tl** thallium 81	207 **Pb** lead 82	209 **Bi** bismuth 83	[209] **Po** polonium 84	[210] **At** astatine 85	[222] **Rn** radon 86
[223] **Fr** francium 87	[226] **Ra** radium 88	[227] **Ac*** actinium 89	[261] **Rf** rutherfordium 104	[262] **Db** dubnium 105	[266] **Sg** seaborgium 106	[264] **Bh** bohrium 107	[277] **Hs** hassium 108	[268] **Mt** meitnerium 109	[271] **Ds** darmstadtium 110	[272] **Rg** roentgenium 111	[285] **Cn** copernicium 112		[286] **Nh** nihonium 113	[289] **Fl** flerovium 114	[289] **Mc** moscovium 115	[293] **Lv** livermorium 116	[294] **Ts** tennessine 117	[294] **Og** oganesson 118

*The lanthanides (atomic numbers 58–71) and the actinides (atomic numbers 90–103) have been omitted.

OXFORD
UNIVERSITY PRESS

Great Clarendon Street, Oxford, OX2 6DP, United Kingdom

Oxford University Press is a department of the University of Oxford.
It furthers the University's objective of excellence in research,
scholarship, and education by publishing worldwide. Oxford is a
registered trade mark of Oxford University Press in the UK and in
certain other countries

British Library Cataloguing in Publication Data
Data available

978-0-19-842382-9

10 9 8 7 6 5 4 3 2 1

Paper used in the production of this book is a natural, recyclable
product made from wood grown in sustainable forests.
The manufacturing process conforms to the environmental regulations
of the country of origin.

Printed in Great Britain by Ashford Colour Press Ltd. Gosport

Acknowledgements

Cover image: Anekoho/Shutterstock; **p32**: Sylvie Bouchard/
Shutterstock; **p87**: namatae/Shutterstock; **p89**: Reddogs/Shutterstock;
p91: tlorna/Shutterstock; **p92**: Pavel Vakhrushev/Shutterstock; **p100**:
PerWil/Shutterstock; **p103**: GIPhotoStock/Science Source

All artwork by Aptara Inc., Q2A Media Services Ltd., and Phoenix
Photosetting